THIRD SUPPLEMENT TO THE THIRD EDITION

FOOD CHEMICALS CODEX

COMMITTEE ON FOOD CHEMICALS CODEX

Food and Nutrition Board
Institute of Medicine
National Academy of Sciences

NATIONAL ACADEMY PRESS

Washington, D.C. 1992

NATIONAL ACADEMY PRESS   **2101 CONSTITUTION AVENUE, NW**   **WASHINGTON, DC 20418**

NOTICE   The project that is the subject of this report was approved by the Governing Board of the National Research Council, whose members are drawn from the Councils of the National Academy of Sciences, the National Academy of Engineering, and the Institute of Medicine. The members of the Committee responsible for the report were chosen for their special competences and with regard for appropriate balance.

INSTITUTE OF MEDICINE   The Institute of Medicine was chartered in 1970 by the National Academy of Sciences to enlist distinguished members of the appropriate professions in the examination of policy matters pertaining to the health of the public. In this, the Institute acts under both the Academy's 1863 congressional charter responsibility to be an adviser to the federal government and its own initiative in identifying issues of medical care, research, and education. Dr. Stuart Bondurant is Acting President of the Institute of Medicine.

FOOD AND NUTRITION BOARD   The Food and Nutrition Board (FNB) was established in 1940 to address issues of national importance that pertain to the safety and adequacy of the nation's food supply; to establish principles and guidelines for adequate nutrition; and to render authoritative judgment on the relationships among food intake, nutrition, and health. The FNB is a multidisciplinary group of scientists with expertise in various aspects of nutrition, nutritional biochemistry, food science and technology, epidemiology, food toxicology, food safety, public health, and food and nutrition policy. These scientists respond to requests from federal agencies about issues concerning food and nutrition, initiate studies that are later assigned to standing or ad hoc FNB committees, and oversee the work of these committees.

Through members of its liaison panels, technical input in aspects of nutrition, food safety, food technology, and food processing is provided.

This study is supported by U.S. Food and Drug Administration Contract No. 223-88-2141.

COMPLIANCE WITH FEDERAL STATUTES   The fact that an article appears in the *Food Chemicals Codex* or its supplements does not exempt it from compliance with requirements of acts of Congress, with regulations and rulings issued by agencies of the United States Government under authority of these acts, or with requirements and regulations of governments in other countries that have adopted the *Food Chemicals Codex*. Revisions of the federal requirements that affect the Codex specifications will be included in Codex supplements as promptly as practicable.

EFFECTIVE DATE   The specifications in this supplement become effective March 1, 1992.

LIBRARY OF CONGRESS CATALOG NUMBER 91-66382
INTERNATIONAL STANDARD BOOK NUMBER 0-309-04585-1

Copyright © 1992 by the National Academy Press

No part of this publication may be reproduced by any mechanical, photographic, or electronic process, or in the form of a phonographic recording, nor may it be stored in a retrieval system, transmitted, or otherwise copied for public or private use, without written permission from the publisher, except for official use by the United States Government or by governments in other countries that have adopted the Food Chemicals Codex.

Printed in the United States of America

Contents

ADDITIONS, CHANGES, AND CORRECTIONS, 95

1 GENERAL PROVISIONS APPLYING TO SPECIFICATIONS, TESTS, AND ASSAYS OF THE *FOOD CHEMICALS CODEX*, 96

2 MONOGRAPHS, 97

3 SPECIFICATIONS FOR FLAVOR AROMATIC CHEMICALS AND ISOLATES, 156

4 TEST METHODS FOR FLAVOR AROMATIC CHEMICALS AND ISOLATES, 166

5 GLC ANALYSIS OF FLAVOR AROMATIC CHEMICALS AND ISOLATES, 167

6 GENERAL TESTS AND APPARATUS, 168

7 SOLUTIONS AND INDICATORS, 173

8 GENERAL INFORMATION, 175

9 INFRARED SPECTRA, 178

INDEX, 181

COMMITTEE ON FOOD CHEMICALS CODEX (1984–1991)

Steve L. Taylor, Department of Food Science and Technology, Food Processing Center, University of Nebraska, Lincoln, NE, *Chair* (from 1989)
Harold T. McNair, Department of Chemistry, Virginia Polytechnic Institute and State University, Blacksburg, VA, *Chair* (1984–1989)
Samuel M. Tuthill, Mallinckrodt, Inc., St. Louis, MO, *Vice-Chair*
Herbert Blumenthal, Silver Spring, MD (from 1989)
Joseph T. Brady, Belleville, IL
Andrew Ebert, The Robert H. Kellen Company, Atlanta, GA
Susan K. Harlander, Department of Food Science and Nutrition, University of Minnesota, St. Paul, MN (from 1989)
Joseph H. Hotchkiss, Institute of Food Science, Department of Food Science, New York State College of Agriculture and Life Sciences, Cornell University, Ithaca, NY (from 1989)
B.L. Huston, Chemical Evaluation Division, Health Protection Branch, Ottawa, Ontario, Canada
Francis P. Mahn, Hoffmann-LaRoche, Inc., Nutley, NJ
Andrew J. Schmitz, Jr., Huntington, NY (from 1989)
Stephen G. Schulman, College of Pharmacy, University of Florida, Gainesville, FL
James T. Stewart, Department of Medicinal Chemistry and Pharmacognosy, College of Pharmacy, The University of Georgia, Athens, GA (1985–1989)
Jan Stofberg, Palmyra, VA

Sanford W. Bigelow, *Project Director* (1989–present)
Durward F. Dodgen, *Project Director* (1988–1989)
Robert A. Mathews, *Project Director* (1984–1987)
Fatima N. Johnson, *Senior Staff Officer*
Sheila A. Mylet, *Research Associate*
Marcia Lewis, *Senior Secretary*

FOOD AND NUTRITION BOARD

M.R.C. Greenwood, *Chair*, Dean of Graduate Studies, University of California, Davis, CA
Donald B. McCormick, *Vice-Chair*, Department of Biochemistry, Emory University School of Medicine, Atlanta, GA
DeWitt S. Goodman, *Vice-Chair*, Institute of Human Nutrition, Columbia University, New York, NY
Edwin L. Bierman, Division of Metabolism, Endocrinology, and Nutrition, University of Washington School of Medicine, Seattle, WA
Edward J. Calabrese, Division of Public Health, University of Massachusetts, Amherst, MA
Johanna T. Dwyer, Frances Stern Nutrition Center, Tufts University, Boston, MA
John W. Erdman, Jr., Division of Nutritional Sciences, University of Illinois, Urbana, IL
Cutberto Garza, Division of Nutritional Sciences, Cornell University, Ithaca, NY
Richard J. Havel, Cardiovascular Research Institute, University of California School of Medicine, San Francisco, CA
Janet C. King, Department of Nutritional Sciences, University of California, Berkeley, CA
John E. Kinsella, School of Agriculture and Environmental Sciences, University of California, Davis, CA
Laurence N. Kolonel, Cancer Center of Hawaii, University of Hawaii, Honolulu, HI
Walter Mertz, Agricultural Research Service, U.S. Department of Agriculture, Beltsville, MD
Malden C. Nesheim, Provost, Cornell University, Ithaca, NY

Catherine E. Woteki, *Executive Director*

Additions, Changes, and Corrections

Additions, changes, and corrections listed herein constitute revisions in the *Food Chemicals Codex*, Third Edition (FCC III). Page numbers refer to FCC III and its first two supplements unless indicated by a reference to pages in THIS SUPPLEMENT.

1/ *General Provisions Applying to Specifications, Tests, and Assays of the* Food Chemicals Codex

Replace the *General Provision* for *Reference Standards* with the following:

Reference Standards Some instrumental and chromatographic tests and assays specify the use of a reference standard. Where a reference standard is designated as FCC or USP, it may be obtained from the United States Pharmacopeia, 12601 Twinbrook Parkway, Rockville, MD 20852. Where a reference standard is designated as an NIST Standard Reference Material, it may be obtained from the National Institute for Standards and Technology, Office of Standard Reference Materials, Room 205, Building 202, Gaithersburg, MD 20899.

To serve its intended purpose, each reference standard must be properly stored, handled, and used. Generally, reference standards should be stored in their original containers away from heat and protected from light. Follow any special instructions accompanying the containers.

Assay and test results are determined on the basis of comparison of the test sample with the reference standard that has been freed from or corrected for volatile residues or water content as instructed on the reference standard label. Where a reference standard is required to be dried before using, transfer a sufficient amount to a clean, dry vessel. Do not use the original container as the drying vessel, and do not dry a reference standard repeatedly at temperatures above 25°. Where the titrimetric determination of water is required at the time a reference standard is to be used, proceed as directed for the *Karl Fischer Titrimetric Method*, page 552.

Unless a reference standard label bears a specific potency or content, the reference standard is taken as being 100.0% pure.

2/ Monographs

N-Acetyl-L-Methionine, page 10

Change the *Molecular Weight* to read:

Mol wt 191.26

Change the *Requirements* entitled *Assay* and *Arsenic* to read:

Assay Not less than 98.5% and not more than 101.5% $C_7H_{13}NO_3S$, calculated on the dried basis.
Arsenic (as As) Not more than 1.5 mg/kg.

Change the last sentence of the *Test* entitled *Assay* to read:

Each mL of 0.1 N iodine is equivalent to 9.563 mg of $C_7H_{13}NO_3S$.

Change the *Test* entitled *Arsenic* to read:

Arsenic A *Sample Solution* prepared as directed for organic compounds meets the requirements of the *Arsenic Test*, page 464, using 1.5 mL of the *Standard Arsenic Solution* in the control (1.5 µg As).

Insert the following new monograph to precede the monograph entitled *Adipic Acid*, page 11:

Aconitic Acid

Equisetic Acid; Citridic Acid; Achilleic Acid

$C_6H_6O_6$ Mol wt 174.11

$C_3H_3(COOH)_3$

DESCRIPTION

Aconitic acid (1-propene-1,2,3-tricarboxylic acid) occurs in the leaves and tubers of *Ranunculaceae Aconitum napellus* L. and various species of *Achillea* and *Equisetum*, in beet root, and in sugar cane. It may be synthesized by the dehydration of citric acid by sulfuric or methanesulfonic acid. Aconitic acid from the above sources has the "*trans*" configuration. It has a melting point of 195° to 200°, with decomposition. It is soluble in water and in alcohol and slightly soluble in ether.

REQUIREMENTS

Identification

The substance as a potassium bromide dispersion exhibits infrared absorption bands at 3030, 2630, and 1720 cm^{-1}. An aqueous solution of the substance exhibits major absorption peaks at 411 and 432 nm with little or no absorption at 389 nm.

Assay Not less than 98.0% $C_3H_3(COOH)_3$.
Arsenic (as As) Not more than 3 mg/kg.
Heavy Metals (as Pb) Not more than 10 mg/kg.
Oxalate Passes test.
Readily Carbonizable Substances Passes test.
Residue on Ignition Not more than 0.1%.
Tridodecylamine Not more than 0.1 mg/kg.
Ultraviolet Absorbance For the specified spectral ranges: 280–289 nm, not more than 0.25 AU; 290–299 nm, not more than 0.20 AU; 300–359 nm, not more than 0.13 AU; 360–400 nm, not more than 0.03 AU.
Water Not more than 0.5%.

TESTS

Assay Proceed as directed under *Assay* in the *Citric Acid* monograph, page 86. Each mL of 1 *N* sodium hydroxide is equivalent to 58.04 mg of $C_3H_3(COOH)_3$.
Arsenic A *Sample Solution* prepared as directed for organic compounds meets the requirements of the *Arsenic Test*, page 464.
Heavy Metals A solution of 2 g in 25 mL of water meets the requirements of the *Heavy Metals Test*, page 512, using 20 μg of lead ion (Pb) in the control (*Solution A*).
Oxalate Neutralize 10 mL of a 1 in 10 solution with ammonia TS, add 5 drops of diluted hydrochloric acid TS, cool, and add 2 mL of calcium chloride TS. No turbidity is produced.
Readily Carbonizable Substances Transfer 1.0 g, finely powdered, to a 22-mm × 175-mm test tube, previously rinsed with 10 mL of sulfuric acid TS and allowed to drain for 10 min. Add 10 mL of sulfuric acid TS, agitate the tube until solution is complete, and immerse the tube in a water bath at 90° ± 1° for 60 ± 0.5 min, keeping the level of the acid below the level of the water during the heating period. Cool the tube in a stream of water, and transfer the acid solution to a color-comparison tube. The color of the acid solution is not darker than that of the same volume of *Matching Fluid K* in a similar matching tube, viewing the tubes vertically against a white background.
Residue on Ignition Ignite 4 g as directed in the general method, page 533.
Tridodecylamine
  Indicator Buffer Solution Prepare a mixture consisting of 700 mL of 0.1 *M* citric acid (anhydrous, reagent grade), 200 mL of 0.2 *M* disodium phosphate, and 50 mL each of 0.2% bromophenol blue and 0.2% bromocresol green in spectrograde methanol.
  No-Indicator Buffer Solution Prepare a mixture consisting of 700 mL of 0.1 *M* citric acid (anhydrous, reagent grade), 200 mL of 0.2 *M* disodium phosphate, and 100 mL of spectrograde methanol.
  Amine Stock Solution Transfer between 40 and 45 mg of tridodecylamine (trilaurylamine), accurately weighed, into a 500-mL volumetric flask, dilute to volume with isopropyl alcohol, and mix. Discard after 3 weeks.
  Standard Amine Solution Prepare the solution fresh daily. Using a graduated 5-mL pipet, transfer into a 100-mL volumetric flask an amount of *Amine Stock Solution* equivalent to 400 μg of tridodecylamine, dilute to volume with isopropyl alcohol, and mix.
  Procedure Dissolve 160 g of anhydrous reagent-grade citric acid (not the sample to be tested) in 320 mL of water, and divide the solution equally between two 250-mL separators, S_1 and S_2. To S_1 add 5 mL of *No-Indicator Buffer Solution*. To S_2 add 2.0 mL of *Standard Amine Solution* and 5 mL of *Indicator Buffer Solution*.

  To prepare solutions of the sample being tested, dissolve 145 g of the anhydrous aconitic acid sample in 320 mL of water. Divide the test solution equally between two 250-mL separators, S_3 and S_4. Add 5 mL of *No-Indicator Buffer Solution* to S_3 and 5 mL of *Indicator Buffer Solution* to S_4.

  To each of the four separators, add 20 mL of a 1:1 mixture (v/v) prepared from spectrograde chloroform and *n*-heptane, shake for 15 min on a mechanical shaker, and allow the phases to separate for 45 min. Drain all except the last few drops of the lower (aqueous) phases and discard. Hand-shake the organic phases with 25 mL each of 0.05 *N* sulfuric acid for 30 s, and allow the phases to separate for 30 min. Drain all except the last few drops of the lower (organic) phase through dry Whatman No. 40 (or equivalent) paper, and collect the aqueous filtrates in separate, small, glass-stoppered containers.

  Determine the absorbance of each solution in a 5-cm cell at 400 nm, with a suitable spectrophotometer standardized before analysis, against chloroform–heptane (1:1 v/v). The net absorbance of the sample ($S_4 - S_3$) is not greater than that of the standard ($S_2 - S_1$).
Ultraviolet Absorbance Determine as directed for *Ultraviolet Absorbance* under *Citric Acid*, page 87.
Water Determine under *Water Determination* by the *Karl Fischer Titrimetric Method*, page 552.

Functional Use in Foods Flavoring substance and adjuvant.
Packaging and Storage Store in tight containers.

DL-Alanine, page 12

Change the *Requirements* entitled *Assay* and *Arsenic* to read:

Assay Not less than 98.5% and not more than 101.5% $C_3H_7NO_2$, calculated on the dried basis.
Arsenic (as As) Not more than 1.5 mg/kg.

Change the *Test* entitled *Arsenic* to read:

Arsenic A *Sample Solution* prepared as directed for organic compounds meets the requirements of the *Arsenic Test*, page 464, using 1.5 mL of the *Standard Arsenic Solution* in the control (1.5 µg As).

Change the *Functional Use in Foods* statement to read:

Functional Use in Foods Nutrient; dietary supplement; flavor enhancer.

L-Alanine, page 13

Change the *Requirements* entitled *Assay* and *Arsenic* to read:

Assay Not less than 98.5% and not more than 101.5% $C_3H_7NO_2$, calculated on the dried basis.
Arsenic (as As) Not more than 1.5 mg/kg.

Change the *Test* entitled *Arsenic* to read:

Arsenic A *Sample Solution* prepared as directed for organic compounds meets the requirements of the *Arsenic Test*, page 464, using 1.5 mL of the *Standard Arsenic Solution* in the control (1.5 µg As).

Aluminum Sodium Sulfate, page 16

Change the *Requirements* entitled *Assay* and *Neutralizing Value* to read:

Assay *Anhydrous form*: not less than 99.0% and not more than 104.0% $AlNa(SO_4)_2$ after drying; *dodecahydrate*: not less than 99.5% $AlNa(SO_4)_2$ after drying.
Neutralizing Value *Anhydrous form*: between 104 and 108.

Annatto Extracts, Second Supplement, page 37

Change the *Requirement* entitled *Residual Solvent* to read:

Residual Solvent *Acetone*: not more than 0.003%; *hexanes*: not more than 0.0025%; *isopropyl alcohol*: not more than 0.005%; *methyl alcohol*: not more than 0.005% in excess of that produced naturally.

Change the *Test* entitled *Color Intensity, Oil-Soluble Extracts*, to read:

Transfer an accurately weighed sample to a solution of 1% glacial acetic acid in acetone, and dilute to a suitable volume (absorbance of 0.5–1.0). Filter the sample to clarify if necessary. Measure the absorbance at 454 nm and calculate the color intensity (I) by the formula:

$$I = A/(b \times c),$$

in which A is the absorbance; b is the cell length, in cm; and c is the concentration, in g/L.

L-Arginine, page 26

Change the *Requirements* entitled *Assay, Arsenic, Loss on Drying*, and *Specific Rotation* to read:

Assay Not less than 98.5% and not more than 101.5% $C_6H_{14}N_4O_2$, calculated on the dried basis.
Arsenic (as As) Not more than 1.5 mg/kg.
Loss on Drying Not more than 1.0%.
Specific Rotation $[\alpha]_D^{20°}$: Between +26.0° and +27.4°, calculated on the dried basis.

Change the *Test* entitled *Arsenic* to read:

Arsenic A *Sample Solution* prepared as directed for organic compounds meets the requirements of the *Arsenic Test*, page 464, using 1.5 mL of the *Standard Arsenic Solution* in the control (1.5 µg As).

L-Arginine Monohydrochloride, page 26

Change the *Requirements* entitled *Assay, Arsenic,* and *Specific Rotation* to read:

Assay Not less than 98.5% and not more than 101.5% $C_6H_{14}N_4O_2 \cdot HCl$, calculated on the dried basis.
Arsenic (as As) Not more than 1.5 mg/kg.
Specific Rotation $[\alpha]_D^{20°}$: Between +21.3° and +23.5°, calculated on the dried basis.

Change the *Test* entitled *Arsenic* to read:

Arsenic A *Sample Solution* prepared as directed for organic compounds meets the requirements of the *Arsenic Test*, page 464, using 1.5 mL of the *Standard Arsenic Solution* in the control (1.5 µg As).

L-Asparagine, page 28

Replace the last sentence of the *Identification* with the following:

The vapor evolved changes the color of acetaldehyde test paper to blue.

Change the *Requirements* entitled *Assay* and *Arsenic* to read:

Assay Not less than 98.0% and not more than 101.5% $C_4H_8N_2O_3$, calculated on the dried basis.
Arsenic (as As) Not more than 1.5 mg/kg.

Insert the following under *Requirements*:

Lead Not more than 5 mg/kg.

Change the *Test* entitled *Arsenic* to read:

Arsenic A *Sample Solution* prepared as directed for organic compounds meets the requirements of the *Arsenic Test*, page 464, using 1.5 mL of the *Standard Arsenic Solution* in the control (1.5 µg As).

Insert the following under *Tests*:

Lead A *Sample Solution* prepared as directed for organic compounds meets the requirements of the *Lead Limit Test*, page 518, using 5 µg of lead ion (Pb) in the control.

DL-Aspartic Acid, page 30

Change the *Requirements* entitled *Assay* and *Arsenic* to read:

Assay Not less than 98.5% and not more than 101.5% $C_4H_7NO_4$, calculated on the dried basis.
Arsenic (as As) Not more than 1.5 mg/kg.

Replace the *Test* entitled *Assay* with the following:

Assay Dissolve about 200 mg, accurately weighed, in 3 mL of formic acid and 50 mL of glacial acetic acid, add 2 drops of crystal violet TS, and titrate with 0.1 N perchloric acid to a green endpoint or until the blue color disappears completely. Each mL of 0.1 N perchloric acid is equivalent to 13.31 mg of $C_4H_7NO_4$.

Change the *Test* entitled *Arsenic* to read:

Arsenic A *Sample Solution* prepared as directed for organic compounds meets the requirements of the *Arsenic Test*, page 464, using 1.5 mL of the *Standard Arsenic Solution* in the control (1.5 µg As).

L-Aspartic Acid, page 30

Change the *Requirements* entitled *Assay*, *Arsenic*, and *Heavy Metals* to read:

Assay Not less than 98.5% and not more than 101.5% $C_4H_7NO_4$, calculated on the dried basis.
Arsenic (as As) Not more than 1.5 mg/kg.
Heavy Metals (as Pb) Not more than 10 mg/kg.

Insert the following under *Requirements*:

Lead Not more than 5 mg/kg.

Replace the *Test* entitled *Assay* with the following:

Assay Dissolve about 200 mg, accurately weighed, in 3 mL of formic acid and 50 mL of glacial acetic acid, add 2 drops of crystal violet TS, and titrate with 0.1 N perchloric acid to a green endpoint or until the blue color disappears completely. Each mL of 0.1 N perchloric acid is equivalent to 13.31 mg of $C_4H_7NO_4$.

Change the *Tests* entitled *Arsenic* and *Heavy Metals* to read:

Arsenic A *Sample Solution* prepared as directed for organic compounds meets the requirements of the *Arsenic Test*, page 464, using 1.5 mL of the *Standard Arsenic Solution* in the control (1.5 µg As).
Heavy Metals Prepare and test a 2-g sample as directed in *Method II* under the *Heavy Metals Test*, page 513, using 20 µg of lead ion (Pb) in the control (*Solution A*).

Insert the following under *Tests*:

Lead A *Sample Solution* prepared as directed for organic compounds meets the requirements of the *Lead Limit Test*, page 518, using 5 µg of lead ion (Pb) in the control.

Black Pepper Oil, page 39

Insert the following under *Requirements*:

Specific Rotation $[\alpha]_D^{20°}$: Between +1° and −33.5°.

Insert the following under *Tests*:

Specific Rotation, page 530 Determine in a solution containing 5 g in sufficient 2 N hydrochloric acid to make 100 mL.

Insert the following new monograph to precede the monograph entitled *Butylated Hydroxymethylphenol*, page 43:

Butane

n-Butane

CH₃CH₂CH₂CH₃   Mol wt 58.12

C₄H₁₀

DESCRIPTION

A colorless, flammable gas with a characteristic odor (boiling temperature is –0.5°). One volume of water dissolves 0.15 volume, and 1 volume of alcohol dissolves 18 volumes at 17° and 770 mm of mercury; 1 volume of ether or chloroform at 17° dissolves 25 or 30 volumes, respectively. Vapor pressure at 21° is about 1620 mm of mercury (17 psi).

REQUIREMENTS

Caution: Butane is highly flammable and explosive. Observe precautions and perform sampling and analytical operations in a well-ventilated fume hood.

Identification

A. The infrared absorption spectrum of Butane exhibits maxima, among others, at about the following wavelengths, in μm: 3.4 (vs), 6.8 (s), 7.2 (m), and 10.4 (m).
B. The vapor pressure of a test specimen, obtained as directed in the *Sampling Procedure* and determined at 21° by means of a suitable pressure gauge, is between 205 and 235 kPa absolute (30 and 34 psia, respectively).

Assay Not less than 97.0% of C₄H₁₀.
Acidity of Residue Passes test.
High-Boiling Residue Not more than 5 mg/kg.
Sulfur Compounds Passes test.
Water Not more than 10 mg/kg.

TESTS

Sampling Procedure Use a stainless steel specimen cylinder equipped with a stainless steel valve and having a capacity of not less than 200 mL and a pressure rating of 240 psi or more. Dry the cylinder with the valve open at 110° for 2 h, and evacuate the hot cylinder to less than 1 mm of mercury. Close the valve, and cool and weigh the cylinder. Tightly connect one end of a charging line to the Butane container, and loosely connect the other end to the specimen cylinder. Carefully open the Butane container, and allow the Butane to flush out the charging line through the loose connection. Avoid excessive flushing that causes moisture to freeze in the charging line and connections. Tighten the fitting on the specimen cylinder, and open the specimen cylinder valve, allowing the Butane to flow into the evacuated cylinder. Continue sampling until the desired amount of specimen is obtained, then close the Butane container valve, and finally, close the specimen cylinder valve. (*Caution*: Do not overload the specimen cylinder.) Again weigh the charged specimen cylinder, and calculate the specimen weight.

Assay
  Chromatographic System Under typical conditions, the gas chromatograph is equipped with a thermal-conductivity detector and contains a 6-m × 3-mm aluminum column packed with 10 weight percent tetraethylene glycol dimethyl ether liquid phase on a support of crushed firebrick (GasChrom R or equivalent), which has been calcined or burned with a clay binder above 900° and silanized. Helium is used as the carrier gas at a flow rate of 50 mL/min, and the temperature of the column is maintained at 33°.
  System Suitability The peak responses obtained for Butane in the chromatograms from duplicate determinations agree within 1%.
  Procedure Connect one Butane cylinder to the chromatograph through a suitable sampling valve and a flow control valve downstream from the sampling valve. Flush the liquid specimen through the sampling valve, taking care to avoid trapping gas or air in the valve. Inject a suitable volume, typically 2 μL, of Butane into the chromatograph, and record the chromatogram. Calculate the percentage purity by dividing 100 times the Butane response by the sum of all the responses in the chromatogram.

Acidity of Residue Add 10 mL of water to the residue obtained in *High-Boiling Residue* (see below), mix by swirling for about 30 s, add 2 drops of methyl orange TS, insert the stopper in the tube, and shake vigorously. No pink or red color appears in the aqueous layer.

High-Boiling Residue Prepare a cooling coil from copper tubing (about 6 mm outside diameter × about 6.1 m long) to fit into a suitable vacuum-jacketed flask. Immerse the cooling coil in a mixture of dry ice and acetone in a vacuum-jacketed flask, and connect one end of the tubing to a specimen cylinder (see *Sampling Procedure*). Carefully open the specimen cylinder valve, flush the cooling coil with about 50 mL of the liquified Butane, and discard this portion of liquid. Continue delivering liquid from the cooling coil, and collect it in a previously chilled 1000-mL sedimentation cone until the cone is filled to the 1000-mL mark (approximately 600 g). Allow the liquid to evaporate, using a warm water bath maintained at about 40° to reduce evaporating time. When all of the liquid has evaporated, rinse the sedimentation cone with two 50-mL portions of pentane, and combine the rinsings in a tared 150-mL evaporating dish. Transfer 100 mL of the pentane solvent to a second tared 150-mL evaporating dish, place both evaporating dishes on a water bath, evaporate to dryness, and heat the dishes in an oven at 100° for 60 min. Cool the dishes in a desiccator, and weigh. Repeat the heating for 15-min periods until successive weighings are within 0.1 mg. The weight of the residue obtained from the specimen is the difference between the weights of the residues in the two evaporating dishes. Calculate the mg/kg of high-boiling residue based on a sample weight of 600 g.

Sulfur Compounds Carefully open the container valve to produce a moderate flow of gas. Do not direct the gas stream toward the face, but deflect a portion of the stream toward the nose. The gas is free from the characteristic odor of sulfur compounds.

Water Determine by the *Karl Fischer Titrimetric Method*, page 552. Proceed, using the following modifications: (a) Provide the closed-system titrating vessel with an opening through which passes a coarse-porosity gas dispersion tube connected to a sampling cylinder. (b) Dilute the reagent with anhydrous methanol to give a water equivalence factor of between 0.2 and 1.0 mg/mL; age this diluted solution for not less than 16 h before standardization. (c) Obtain a 100-g specimen as directed in the *Sampling Procedure*, and introduce the specimen into the titration vessel through the gas dispersion tube at a rate of about 100 mL of gas per min; if necessary, heat the specimen cylinder gently to maintain this flow rate.

Functional Use in Foods Propellant; aerating agent.
Packaging and Storage Store in tight cylinders protected from excessive heat.

Calcium Chloride, page 47

Change the *Requirement* entitled *Assay* to read:

Assay Not less than 96.0% and not more than 107.0% $CaCl_2 \cdot 2H_2O$.

Calcium Gluconate, page 51

Insert the following formula and molecular weight:

$C_{12}H_{22}CaO_{14} \cdot H_2O$ Mol wt 448.39 (monohydrate)

Insert the following sentence at the beginning of the *Description*:

Calcium gluconate is anhydrous or contains one molecule of water of hydration.

Change the second sentence of the *Description* to read:

It is in the form of white, crystalline granules or a powder.

Replace *Identification Test B* with the following:

Dissolve a quantity of the sample in water to obtain a test solution containing 10 mg/mL, heating in a water bath at 60° if necessary. Similarly, prepare a standard solution of USP Potassium Gluconate Reference Standard in water containing 10 mg/mL. Apply separate 5-µL portions of the test solution and the standard solution on a suitable thin-layer chromatographic plate (see *Thin-Layer Chromatography*, page 474) coated with a 0.25-mm layer of chromatographic silica gel, and allow to dry. Develop the chromatogram in a solvent system consisting of a mixture of alcohol, water, ammonium hydroxide, and ethyl acetate (50:30:10:10) until the solvent front has moved about three-fourths of the length of the plate. Remove the plate from the chamber, and dry at 110° for 20 min. Allow to cool, and spray with a spray reagent prepared as follows: dissolve 2.5 g of ammonium molybdate in about 50 mL of 2 N sulfuric acid in a 100-mL volumetric flask, add 1.0 g of ceric sulfate, swirl to dissolve, dilute with 2 N sulfuric acid to volume, and mix. Heat the plate at 110° for about 10 min: the principal spot obtained from the test solution corresponds in color, size, and R_f value to that obtained from the standard solution.

Replace the *Requirement* entitled *Assay* with the following:

Assay *Anhydrous form*: not less than 98.0% and not more than 102.0% $C_{12}H_{22}CaO_{14}$, calculated on the dried basis; *monohydrate*: not less than 98.0% and not more than 102.0% $C_{12}H_{22}CaO_{14} \cdot H_2O$, calculated on the as-is basis.

Change the *Requirements* entitled *Heavy Metals*, *Lead*, and *Loss on Drying* to read:

Heavy Metals (as Pb) Not more than 10 mg/kg.
Lead Not more than 5 mg/kg.
Loss on Drying *Anhydrous form*: not more than 3.0%; *monohydrate*: not more than 2.0%.

Insert the following heading after the *Requirements* section:

ADDITIONAL REQUIREMENTS

Insert the following *Additional Requirement*:

Labeling Label to indicate whether it is anhydrous or the monohydrate.

Change the first and last sentences of the *Test* entitled *Assay* to read, respectively:

Dissolve about 800 mg, accurately weighed, in 100 mL of water containing 2 mL of diluted hydrochloric acid TS.

Each mL of 0.05 M disodium EDTA is equivalent to 21.52 mg of $C_{12}H_{22}CaO_{14}$ or 22.42 mg of $C_{12}H_{22}CaO_{14} \cdot H_2O$.

Delete the first sentence of the *Test* entitled *Heavy Metals*.

Change the second sentence of the *Test* entitled *Heavy Metals* to read:

Prepare and test a 2-g sample as directed in *Method II* under the *Heavy Metals Test*, page 513, using 20 μg of lead ion (Pb) in the control (*Solution A*).

Change the *Test* entitled *Lead* to read:

Lead A *Sample Solution* prepared as directed for organic compounds meets the requirements of the *Lead Limit Test*, page 518, using 5 μg of lead ion (Pb) in the control.

Replace the *Functional Use in Foods* statement with the following:

Functional Use in Foods Firming agent; formulation aid; sequestrant; stabilizer; thickener; texturizer.

Calcium Pantothenate, page 56

Change the *Requirement* entitled *Assay* to read:

Assay Not less than 97.0% and not more than 103.0% dextrorotatory calcium pantothenate ($C_{18}H_{32}CaN_2O_{10}$), calculated on the dried basis.

Replace the *Test* entitled *Assay* with the following:

Assay (Use low-actinic glassware throughout this procedure.)
Mobile Phase Transfer 4.7 mL of phosphoric acid into a 2-L volumetric flask, and dilute to volume with water. Filter the solution through a 0.45-μm pore-size disk.
Standard Preparation Transfer about 16 mg of USP Calcium Pantothenate Reference Standard, accurately weighed, into a 100-mL volumetric flask. Dilute to volume with *Mobile Phase*, and mix.
Assay Preparation Proceed as directed for the *Standard Preparation*, using an accurately weighed amount of the sample equivalent to about 16 mg of Calcium Pantothenate.

Chromatographic System (see *Chromatography*, page 476) Use a high-pressure liquid chromatograph equipped with an ultraviolet detector that measures at 200 nm. Under typical conditions, the instrument contains a 110- × 4.6-mm cartridge packed with octadecylsilanized silica (5-μm Partisil ODS-3 or equivalent). The flow rate of the *Mobile Phase* is about 2 mL/min.
System Suitability Three replicate injections of the *Standard Preparation* show a relative standard deviation of not more than 2.0% for the response factor of the main peak obtained by the formula (A_s/C_s), in which A_s is the peak area response of the *Standard Preparation*, and C_s is the concentration, in mg/mL, of USP Calcium Pantothenate Reference Standard in the *Standard Preparation*.
Procedure Separately inject about 20 μL of the *Standard Preparation* and the *Assay Preparation* into the chromatograph, and record the chromatograms. Measure the peak area responses of the major peaks. Calculate the quantity, in mg, of $C_{18}H_{32}CaN_2O_{10}$ in the portion of Calcium Pantothenate taken by the formula:

$$100C_s(A_u/A_s),$$

in which C_s is the concentration, in mg/mL, of USP Calcium Pantothenate Reference Standard in the *Standard Preparation*, and A_u and A_s are the peak area responses of the *Assay Preparation* and the *Standard Preparation*, respectively.

Calcium Pantothenate, Calcium Chloride Double Salt, page 57

Change the *Requirement* entitled *Assay* to read:

Assay Not less than 45.0% and not more than 55.0% Calcium Pantothenate ($C_{18}H_{32}Ca_2N_2O_{10}Cl_4$), calculated on the dried basis.

Change the *Test* entitled *Assay* to read:

Assay Proceed as directed for *Assay* under *Calcium Pantothenate*, page 56.

Calcium Pantothenate, Racemic, page 57

Change the *Requirement* entitled *Assay* to read:

Assay Not less than 97.0% and not more than 103.0% Calcium Pantothenate ($C_{18}H_{32}CaN_2O_{10}$), calculated on the dried basis.

Insert the following under *Requirements*:

Specific Rotation $[\alpha]_D^{25°}$: Between –0.05° and +0.05°.

Replace the Test entitled Assay with the following:

Assay Proceed as directed for *Assay* under *Calcium Pantothenate*, page 56.

Insert the following under Tests:

Specific Rotation, page 530 Determine in a solution containing 500 mg, calculated on the dried basis, in each 10-mL portion.

Insert the following new monograph to precede the monograph entitled Calcium Stearate, page 64:

Calcium Sorbate

2,4-Hexadienoic Acid, Calcium Salt

$$(CH_3CH{=}CHCH{=}CHCOO)_2Ca$$

$C_{12}H_{14}CaO_4$ Formula wt 262.31

DESCRIPTION

White, fine crystalline powder. It decomposes at about 400°. It is sparingly soluble in water and practically insoluble in organic solvents as well as in fats and in oils.

REQUIREMENTS

Identification

A. Ignite 1 g at 800°. Cool, and slake with 10 mL of water. Add acetic acid until the sample is dissolved, and filter if necessary. The resultant solution gives positive tests for *Calcium*, page 516.
B. Place 200 mg of the sample in 5 mL of methanol. Add 0.1 mL of sodium hydroxide TS, and dissolve in 95 mL of water. After addition of a few drops of bromine TS, the color is discharged.

Assay Not less than 98.0% and not more than 101.0% $C_{12}H_{14}CaO_4$, calculated on the dried basis.
Acidity (as sorbic acid) Passes test (approximately 1%).
Alkalinity (as Ca(OH)$_2$) Passes test (approximately 0.5%).
Arsenic (as As) Not more than 3 mg/kg.
Heavy Metals (as Pb) Not more than 10 mg/kg.
Lead Not more than 5 mg/kg.
Loss on Drying Not more than 1.0%.

TESTS

Assay Dissolve about 150 mg of the sample, accurately weighed, in 50 mL of glacial acetic acid in a 250-mL glass-stoppered Erlenmeyer flask, warming if necessary to effect solution. Cool to room temperature, add 2 drops of crystal violet TS, and titrate with 0.1 N perchloric acid in glacial acetic acid to a blue-green endpoint that persists for at least 30 s. Perform a blank determination (see page 2), and make any necessary correction. Two mL of 0.1 N perchloric acid is equivalent to 26.23 mg of $C_{12}H_{14}CaO_4$.
Acidity or Alkalinity Add some drops of methanol to 1 g of the sample. Add 30 mL of water and several drops of phenolphthalein TS. If the mixture is colorless, titrate with 0.1 N sodium hydroxide to a pink color that persists for 15 s. Not more than 1.0 mL is required. If the mixture is pink, titrate with 0.1 N hydrochloric acid. Not more than 1.35 mL is required to discharge the pink color.
Arsenic A *Sample Solution* prepared as directed for organic compounds meets the requirements of the *Arsenic Test*, page 464.
Heavy Metals Prepare and test a 2-g sample as directed in *Method II* under the *Heavy Metals Test*, page 513, using 20 µg of lead ion (Pb) in the control (*Solution A*).
Lead Prepare and test a 2-g sample as directed under the *Lead Limit Test*, page 518, using 10 µg of lead ion (Pb) in the control.
Loss on Drying, page 518 Dry at 105° for 3 h.

Functional Use in Foods Preservative.
Packaging and Storage Store in tight containers.

Insert the following new monograph to precede the monograph entitled Canthaxanthin, page 67:

Canola Oil

Low Erucic Acid Rapeseed Oil; LEAR

DESCRIPTION

A light yellow oil, typically obtained by mechanical expression or by *n*-hexane extraction, from the seed of the plant *Brassica napus* or *Brassica campestris* of the family *Cruciferae*. The plant varieties are those producing oil-bearing seeds with a low erucic acid ($C_{22:1}$) content. It is a mixture of triglycerides composed of both saturated and unsaturated fatty acids. It is refined, bleached, and deodorized to substantially remove free fatty acids; phospholipids; color; odor and flavor components; and miscellaneous, other non-oil materials. It can be hydrogenated to reduce the level of unsaturated fatty acids for functional purposes in foods. It is a liquid at 0° and above.

REQUIREMENTS

Identification

Unhydrogenated Canola Oil exhibits the following composition profile of fatty acids as determined under *Fatty Acid Composition*, page 82 of the Second Supplement.

Fatty Acid:	<14	14:0	16:0	16:1	18:0	18:1	18:2
Weight % (Range):	<0.1	<0.2	<6.0	<1.0	<2.5	>50	<40.0
Fatty Acid:	18:3	20:0	20:1	22:0	22:1	24:0	24:1
Weight % (Range):	<14	<1.0	<2.0	<0.5	<2.0	<0.2	<0.2

Acid Value   Not more than 6.
Arsenic (as As)   Not more than 0.5 mg/kg.
Cold Test   Passes test.
Color (Lovibond)   Not more than 1.5 red/15 yellow.
Erucic Acid   Not more than 2.0%.
Free Fatty Acids (as oleic acid)   Not more than 0.05%.
Heavy Metals (as Pb)   Not more than 5 mg/kg.
Iodine Value   Between 110 and 126.
Lead   Not more than 0.1 mg/kg.
Linolenic Acid   Not more than 15.0%.
Peroxide Value   Not more than 10 meq/kg.
Refractive Index $[\alpha]_D^{40°}$: Between 1.465 and 1.467.
Saponifiable Value   Between 178 and 193.
Stability   Not less than 7 h.
Sulfur   Not more than 10 mg/kg.
Unsaponifiable Matter   Not more than 1.5%.
Water   Not more than 0.1%.

ADDITIONAL REQUIREMENTS

Labeling   Hydrogenated Canola Oil less than fully hydrogenated must be labeled as partially hydrogenated Canola Oil.

TESTS

Acid Value   Determine as directed under *Method II* in the general procedure, page 504.
Arsenic   A *Sample Solution* containing 2.0 g of the sample prepared as directed for organic compounds meets the requirements of the *Arsenic Test*, page 464, using 1 mL of the *Standard Arsenic Solution* in the control (1 µg As).
Cold Test   Proceed as directed under *Cold Test*, page 82 of the Second Supplement.
Color   Proceed as directed under *Color*, page 82 of the Second Supplement. Use a 133.4-mm cell.
Erucic Acid   Determine as part of *Fatty Acid Composition*, page 82 of the Second Supplement.
Free Fatty Acids   Proceed as directed under *Free Fatty Acids*, page 504, using the following equivalence factor (*e*) in the formula given in the procedure:

Free fatty acids as oleic acid, *e* = 28.2

Heavy Metals   Prepare and test a 2-g sample as directed in *Method II* under the *Heavy Metals Test*, page 513, using 10 µg of lead ion (Pb) in the control (*Solution A*).
Iodine Value   Proceed as directed under *Wijs Method*, page 505.
Lead   Determine as directed under *Method II* in the *Atomic Absorption Spectrophotometric Graphite Furnace Method* under the *Lead Limit Test*, page 168 of THIS SUPPLEMENT, using a 10-g sample.
Linolenic Acid   Proceed as directed under *Fatty Acid Composition*, page 82 of the Second Supplement.
Peroxide Value   Proceed as directed under *Peroxide Value*, page 148 of the monograph for *Hydroxylated Lecithin*. However, after the addition of saturated potassium iodide and mixing, mix the solution for only 1 min and begin the titration immediately instead of allowing the solution to stand for 10 min.
Refractive Index, page 533   Determine with an Abbé or other refractometer of equal or greater accuracy.
Saponifiable Value   Determine as directed under the general method, page 509.
Stability   Proceed as directed under *Stability*, page 83 of the Second Supplement.
Sulfur   Organosulfur compounds present in the sample react with Raney nickel to produce nickel sulfides. Nickel sulfides are treated with a strong acid to produce hydrogen sulfide, which is trapped and titrated with mercuric acetate using a dithizone indicator. (*Caution*: This test requires the use of the following hazardous substances: mercuric acetate, spongy nickel, and dibenzyl disulfide. Conduct the test in a fume hood.)

Apparatus   Fit a 125-mL round-bottom boiling flask with a cylindrical filling funnel (20 mL with open top), an ST PTFE metering valve stopcock, and a gas inlet tube (see figure 46 on page 171 of THIS SUPPLEMENT). On top of the boiling flask, fit a water-jacketed distillation column with hooks. To the distillation column, fit a piece of glass tubing with ground ST inner joints with hooks, and connect the distillation column and a gas dispersion tube with ST outer joints with hooks.

Dibenzyl Disulfide Solution   Accurately weigh 0.75 g of dibenzyl disulfide and place in a 250-mL volumetric flask. Dilute to volume with methyl isobutyl ketone, and mix.

Sulfur Standard   Accurately weigh five 250.0-g samples of food-grade peanut oil. Transfer 0.0, 1.0, 2.0, 3.0, and 4.0 mL of the *Dibenzyl Disulfide Solution* into the peanut oil samples; the samples contain 0, 3, 6, 9, and 12 mg/kg sulfur, respectively.

Raney Nickel Preparation (*Caution*: Raney nickel is pyrophoric when dry.) Raney nickel is produced by reacting nickel–aluminum alloy with sodium hydroxide. Weigh accurately 1 g of nickel–aluminum alloy powder (50% Ni, 50% Al), place it in a 50-mL centrifuge tube, and chill it in an ice bath. Each pellet is enough catalyst for one determination. Slowly add 5 mL of water per tube, and let the tube stand for 10 min. Then, slowly add 10 mL of 2.5 N sodium hydroxide, and allow the mixture to react for 30 min. Cap the tubes, and place them in a 50° water bath for 2 h. Centrifuge the mixture at 1000 rpm for 10 min, and discard the supernatant. Wash the pellets twice with 15 mL of water and twice with 15 mL of isopropanol, centrifuging between each wash. The catalyst may be stored under isopropanol for a period no longer than 2 weeks.

NOTE: Properly dispose of the unused *Raney Nickel Preparation* by transferring it to a 250-mL Erlenmeyer flask, and place it in a fume hood. Add 20 mL of 60% (w/v) hydrochloric acid, and allow complete digestion of the catalyst. (*Caution*: Hydrogen gas is evolved during the digestion process.)

Dithizone Indicator Solution Dissolve 10 mg of dithizone (diphenylthiocarbazone) in a 10-mL volumetric flask with acetone.

Mercuric Acetate Titrant (NOTE: Mercuric acetate is a strong irritant when ingested or inhaled or upon dermal exposure.) Transfer 3.82 g of mercuric acetate into a 1000-mL volumetric flask containing 950 mL of water. Add 12.2 mL of acetic acid, dilute to volume with water, and mix. Transfer 10.0 mL of this solution into a 100-mL volumetric flask, dilute to volume with water, and mix. The titrant solution contains 0.0012 M mercuric acetate.

Titration Reagent Blank Add 50.0 mL of 1 N sodium hydroxide and 50.0 mL of acetone to a 250-mL beaker, and mix. Add 0.5 mL of the *Dithizone Indicator Solution*, and titrate with *Mercuric Acetate Titrant* until the color changes from bright amber to strawberry red. Record the volume of titrant used.

Procedure Test a representative portion of the sample. Accurately weigh 15 to 20 g of the sample, and place it on the bottom of the boiling flask. Discard the isopropanol from the *Raney Nickel Preparation*, add 10 mL of 95% isopropanol, mix, and place the mixture in the sample oil. Attach the water condenser and the nitrogen line to the boiling flask and adjust the gas flow to 4 psi through the sample. Place a heating mantle under the flask. Immerse the bubbler in a 250-mL beaker containing 50.0 mL of 1 N sodium hydroxide, and stir slowly. Boil the sample for 90 min. Add 50 mL of acetone and 0.5 mL of *Dithizone Indicator Solution* to the 250-mL beaker. Add 20 mL of 60% hydrochloric acid into the filling funnel fitted onto the boiling flask, and adjust the nitrogen flow to 2 to 3 psi. Position the stir bar directly under the bubbler for maximum dispersion of the hydrogen sulfide bubbles. Slowly add the solution of 60% hydrochloric acid to the boiling flask. Begin the titration with *Mercuric Acetate Titrant* until the bright amber color changes to strawberry red. Add enough hydrochloric acid to turn the solution in the boiling flask green, and then let it boil for 15 min. Continue the titration throughout the boiling stage, making sure to rinse the inside of the bubbler with the solution in the beaker by turning off the nitrogen flow until the solution rises to the top of the vertical tube. Rinse the tube a second time (the solution usually returns to amber during the first rinse). Continue the titration and record the volume of titrant used to the nearest 0.01 mL.

Calculation The concentration of sulfur in the sample, in mg/kg, is calculated by the following formula:

$$(V_s - V_b) \times K/W,$$

in which V_s is the volume, in mL, of titrant to the endpoint for the sample; V_b is the volume, in mL, of titrant to the endpoint for the blank (usually about 0.10 mL); K is a constant determined from the calibration of the *Sulfur Standard* (expressed as µg sulfur/mL titrant); and W is the weight, in g, of the sample.

The *Sulfur Standards* are analyzed, in duplicate, to determine the constant, K, and are calculated by the following formula:

$$K = W \times C/(V_s - V_b),$$

where W is the weight, in g, of the *Sulfur Standard*; C is the concentration, in mg/kg, of the *Sulfur Standard*; V_s is the volume, in mL, of titrant for the *Sulfur Standard*; and V_b is the volume, in mL, of titrant for the *Titration Reagent Blank*.

Unsaponifiable Matter Proceed as directed under *Unsaponifiable Matter*, page 509.

Water Proceed as directed under *Water Determination* using the *Karl Fischer Titrimetric Method*, page 553. However, in place of 35–40 mL of methanol, use 50 mL of a 1:1 chloroform–methanol mixture to dissolve the sample.

Functional Use in Foods Coating agent; emulsifying agent; formulation aid; texturizer.

Packaging and Storage Store in well-closed containers.

Carbon, Activated, page 70

Delete the *Synonyms* Decolorizing Carbon; Active Carbon.

Replace the penultimate sentence of the *Description* with the following:

Activated Carbon occurs as a black, tasteless substance, varying in particle size from coarse granules to a fine powder.

Insert the following under *Requirements*:

Iodine Value Not less than 400.

Insert the following under *Tests*:

Iodine Value

Standard Iodine Solution Using a wide-mouth funnel, transfer 805 g of potassium iodide to a 2-L volumetric flask, and add enough water to the flask to cover the sample. Prepare a solution of 50 g of potassium iodide in 150 mL of water in a 250-mL beaker. Weigh 120 g of molecular iodine in a glass-stoppered weighing bottle. Pour the iodine into the funnel fitted to the 2-L volumetric flask, and then immediately restopper the weighing bottle, add an additional 285 g of potassium iodide to the funnel, and wash the funnel clean with a stream of water. With the potassium iodide solution, rinse the remaining iodine from the weighing bottle until the washings are colorless. Pour the remaining potassium iodide solution into the volumetric flask, and rinse all glassware with water into the volumetric flask. Grease the stopper, and insert it into the flask.

Gently shake the volumetric flask on a mechanical shaker for 30 min, add about 300 mL of water, and repeat the shaking until the flask has cooled to room temperature. Add progressively smaller amounts of water to the flask so that the final quantity required to dilute to volume is small enough that no further heat of solution is detectable. Allow the solution to stand overnight.

Transfer 100 mL of the iodine solution to a 1-L volumetric flask, and fill with water to the bottom of the flask neck. Dilute to volume with water after 1 h and mix thoroughly. Allow the solution to stand 30 min before standardizing.

Standardization Transfer approximately 125 mg of primary standard-grade barium thiosulfate (dried at 40°), accurately weighed, to a 125-mL Erlenmeyer flask. Cover the sample with about 50 mL of water, and titrate with iodine solution until the standard has completely dissolved. Add 1 mL of starch TS, and continue the titration until one drop of iodine solution produces a distinct, light blue color. The normality (N) of the iodine solution, which is approximately 0.047, is given by $V \times 0.4673$, in which V is the volume, in mL, of titrant.

Procedure Transfer approximately 50 to 60 mg of sample, previously dried at 105° for 30 min and accurately weighed, to a glass vial or bottle. Pipet 25 mL of *Standard Iodine Solution*, stopper the container, and shake mechanically at about 240 strokes per min for 2 min. Transfer the mixture to a centrifuge tube, and centrifuge until the sample forms a pellet firm enough to permit decanting of the supernatant solution. Pipet 20 mL of the supernatant solution into a 250-mL Erlenmeyer flask, and titrate with sodium thiosulfate (*Volumetric Solution*, p. 567), diluted 1 in 2.5 with water, until the yellow iodine color becomes pale. Add 1 mL of starch TS, and continue titrating until the blue color is discharged. Record the volume of titrant as S. Titrate a 25-mL aliquot of *Standard Iodine Solution* with the sodium thiosulfate, and record the volume of titrant required as B.

Calculation Calculate the iodine value (I), in mg of iodine adsorbed per g of Carbon, by the formula:

$$I = [N(B - S)(126.91)]/W,$$

in which $(B - S)$ is the difference in volumes of sodium thiosulfate required for the blank and the sample, respectively; N is the exact normality of the sodium thiosulfate; and W is the weight, in g, of the sample.

Carnauba Wax, page 73

Change the *Requirements* entitled *Ester Value* and *Melting Range* to read:

Ester Value Between 70 and 80.
Melting Range Between 80° and 86°.

Insert the following under *Requirements*:

Residue on Ignition Not more than 0.25%.
Saponification Value Between 78 and 95.

Delete the first sentence of the *Test* entitled *Ester Value*.

Insert the following under *Tests*:

Residue on Ignition Heat a 2-g sample in a tared, open, porcelain or platinum dish over an open flame. It volatilizes without emitting an acrid odor. Ignite as directed in the general procedure, page 533.
Saponification Value Weigh accurately about 5 g of the sample, and determine as directed in the general procedure, page 509.

Carrot Seed Oil, page 76

Change the *Requirement* entitled *Refractive Index* to read:

Refractive Index Between 1.483 and 1.493 at 20°.

Citric Acid, page 86

Change the *Requirements* entitled *Arsenic*, *Heavy Metals*, and *Ultraviolet Absorbance* to read:

Arsenic (as As) Not more than 1 mg/kg.
Heavy Metals (as Pb) Not more than 5 mg/kg.

Ultraviolet Absorbance (polycyclic aromatic hydrocarbons) 280–289 nm, not more than 0.25; 290–299 nm, not more than 0.20; 300–359 nm, not more than 0.13; 360–400 nm, not more than 0.03.

Insert the following under *Requirements*:

Lead Not more than 0.5 mg/kg.

Insert the following heading after the *Requirements* section:

ADDITIONAL REQUIREMENTS

Insert the following *Additional Requirement* for *Labeling*:

Labeling Label to indicate whether it is anhydrous or hydrous.

Change the *Test* entitled *Arsenic* to read:

Arsenic A *Sample Solution* prepared as directed for organic compounds meets the requirements of the *Arsenic Test*, page 464, using 1 mL of the *Standard Arsenic Solution* in the control (1 μg As).

Insert the following under *Tests*:

Lead Determine as directed under *Method I* in the *Atomic Absorption Spectrophotometric Graphite Furnace Method* under the *Lead Limit Test*, page 168 of THIS SUPPLEMENT, using a 5-g sample.

Replace the first sentence of the second paragraph of the test entitled *Tridodecylamine, Procedure*, with the following:

To prepare solutions of the sample being tested, dissolve 160 g of anhydrous Citric Acid sample in 320 mL of water (or 174 g of Citric Acid monohydrate sample in 306 mL of water).

Insert the following new monograph to precede the monograph entitled *Coriander Oil*, page 90:

Copper Sulfate

Cupric Sulfate

$CuSO_4 \cdot 5H_2O$ Formula wt 249.68

DESCRIPTION

Blue crystals, crystalline granules, or powder. It effloresces slowly in dry air and is freely soluble in water, soluble in glycerin, and slightly soluble in alcohol.

REQUIREMENTS

Identification

A 1 in 20 solution gives positive tests for *Copper*, page 516, and for *Sulfate*, page 517.

Assay Not less than 98.0% and not more than 102.0% $CuSO_4 \cdot 5H_2O$.
Arsenic (as As) Not more than 3 mg/kg.
Iron Not more than 0.01%.
Lead Not more than 10 mg/kg.
Substances Not Precipitated by Hydrogen Sulfide Not more than 0.3%.

TESTS

Assay Dissolve about 1 g of the sample, accurately weighed, in 50 mL of water, add 4 mL of glacial acetic acid and 3 g of potassium iodide, mix well, and titrate with 0.1 *N* sodium thiosulfate to a light yellow color. Add 2 g of ammonium thiocyanate, mix, and then add 3 mL of starch TS, and continue titrating to a milky white endpoint. Perform a blank titration, and make any necessary correction. Each mL of 0.1 *N* sodium thiosulfate is equivalent to 24.97 mg of $CuSO_4 \cdot 5H_2O$.

Arsenic, page 464 Transfer 1 g, accurately weighed, to an arsenic generator flask, and dissolve in 30 mL of water. Add 20 mL of hydrochloric acid containing 3 g of stannous chloride dihydrate (omit the addition of potassium iodide TS). Add 3 g of zinc shot (instead of 20-mesh zinc) to control the reaction, connect the scrubber-absorber assembly, and allow the hydrogen evolution and color development to proceed for 45 min, while maintaining the generator flask at 25° ± 3°. The absorbance of the red color in the absorber solution does not exceed that produced by 3 μg of arsenic treated in a similar manner and held at 40° ± 3° during the hydrogen evolution step. (NOTE: The different temperatures are specified to approximately equalize the rate of hydrogen evolution from the sample and the standard. If necessary, the temperature of the standard may be adjusted to equalize the rates.)

Iron To the residue from the test entitled *Substances Not Precipitated by Hydrogen Sulfide*, add 2 mL of hydrochloric acid and 0.1 mL of nitric acid, cover with a watch glass, and digest on a steam bath for 20 min. Remove the watch glass, and evaporate to dryness. Dissolve the residue in 1 mL of hydrochloric acid, and dilute to 60 mL. Dilute 5 mL of this solution to 40 mL, add 2 mL of hydrochloric acid, and dilute to 50 mL. Add 40 mg of ammonium peroxydisulfate crystals and 10 mL of ammonium thiocyanate TS, and mix thoroughly. Any red color produced within 1 h shall not exceed that produced by 0.033 mg of Fe in an equal volume of solution containing the reagents used in the test.

Lead A solution of 1 g in 25 mL of water meets the requirements of the *Lead Limit Test*, page 518, using 10 μg of lead ion (Pb) in the control.

Substances Not Precipitated by Hydrogen Sulfide Dissolve 5 g in 200 mL of sulfuric acid (1 in 100), heat to 70°, and pass hydrogen sulfide through the solution until the copper is completely precipitated. Dilute to 250 mL, mix thoroughly, allow the precipitate to settle, and filter. Evaporate 200 mL of the filtrate to dryness in a tared dish, ignite at $800° \pm 25°$ for 15 min, cool, and weigh.

Functional Use in Foods Nutrient supplement; processing aid.

Packaging and Storage Store in tight containers.

L-Cysteine Monohydrochloride, page 92

Replace the *Requirements* entitled *Assay* and *Arsenic* to read:

Assay Not less than 98.0% and not more than 101.5% $C_3H_7NO_2S \cdot HCl$, calculated on the dried basis.

Arsenic (as As) Not more than 1.5 mg/kg.

Change the *Test* entitled *Arsenic* to read:

Arsenic A *Sample Solution* prepared as directed for organic compounds meets the requirements of the *Arsenic Test*, page 464, using 1.5 mL of the *Standard Arsenic Solution* in the control (1.5 μg As).

L-Cystine, page 93

Insert the following under *Identification*:

Identification The infrared absorption spectrum of a potassium bromide dispersion of L-Cystine, previously dried, exhibits maxima only at the same wavelengths as those of a similar preparation of NIST Cystine Standard Reference Material.

Replace the *Requirements* entitled *Assay*, *Arsenic*, and *Heavy Metals* to read:

Assay Not less than 98.5% and not more than 101.5% $C_6H_{12}N_2O_4S_2$, calculated on the dried basis.

Arsenic (as As) Not more than 1.5 mg/kg.

Heavy Metals (as Pb) Not more than 20 mg/kg.

Delete the *Requirements* entitled *Iron* and *Nitrogen*.

Replace the *Test* entitled *Assay* to read:

Assay Determine as directed under *Nitrogen Determination*, page 521, using a 200-mg sample. Percent L-Cystine equals percent $N \times 8.58$.

Change the *Tests* entitled *Arsenic* and *Heavy Metals* to read:

Arsenic A *Sample Solution* prepared as directed for organic compounds meets the requirements of the *Arsenic Test*, page 464, using 1.5 mL of the *Standard Arsenic Solution* in the control (1.5 μg As).

Heavy Metals Prepare and test a 1-g sample as directed in *Method II* under the *Heavy Metals Test*, page 513, using 20 μg of lead ion (Pb) in the control (*Solution A*).

Delete the *Tests* entitled *Iron* and *Nitrogen*.

Dextrose, page 97

Replace the *Requirements* entitled *Arsenic*, *Heavy Metals*, *Loss on Drying*, *Residue on Ignition*, and *Specific Rotation* to read:

Arsenic (as As) Not more than 1 mg/kg.

Heavy Metals (as Pb) Not more than 5 mg/kg.

Loss on Drying *Anhydrous form*: not more than 2.0%; *monohydrate*: not more than 10.0%.

Residue on Ignition Not more than 0.1%.

Specific Rotation $[\alpha]_D^{25°}$: Between +52.6° and +53.2° after drying.

Insert the following under *Requirements*:

Lead Not more than 0.1 mg/kg.

Replace the *Tests* entitled *Assay*, *Arsenic*, *Heavy Metals*, and *Residue on Ignition*, to read:

Assay Determine as directed under the *Reducing Sugars Assay*, page 169 of THIS SUPPLEMENT.

Arsenic A *Sample Solution* prepared using a 1-g sample meets the requirements of the *Arsenic Test*, page 464, using 1 mL of *Standard Arsenic Solution* in the control (1 μg As).

Heavy Metals Prepare and test a 4-g sample as directed under the *Heavy Metals Limit Test*, page 512, using 20 μg of lead (Pb) in the control (*Solution A*).

Residue on Ignition Ignite 10 g as directed in the general method, page 533.

Change the *Test* entitled *Sulfur Dioxide* to read:

Sulfur Dioxide Determine as directed in the general method, page 170 of THIS SUPPLEMENT, using a 75-g sample.

Insert the following under *Tests*:

Lead Determine as directed under *Method I* in the *Atomic Absorption Spectrophotometric Graphite Furnace Method* under the *Lead Limit Test*, page 000 of THIS SUPPLEMENT, using a 5-g sample.

Diatomaceous Earth, page 99

Change the *Requirements* entitled *Loss on Drying*, *Loss on Ignition*, and *Nonsiliceous Substances* to read:

Loss on Drying *Natural powders*: not more than 10.0%; *calcined* and *flux-calcined powders*: not more than 3.0%.
Loss on Ignition *Natural powders*: not more than 7.0%, calculated on the dried basis; *calcined* and *flux-calcined powders*: not more than 0.5%, calculated on the dried basis.
Nonsiliceous Substances Not more than 25.0%, calculated on the dried basis.

Change the *Test* entitled *Loss on Ignition* to read:

Loss on Ignition Weigh accurately about 1 g, and ignite at 800° to weight in a suitable tared crucible.

Ethoxyquin, page 112

Change *Functional Use in Foods* to read:

Functional Use in Foods Antioxidant.

Fennel Oil, page 115

Change the *Requirement* entitled *Refractive Index* to read:

Refractive Index Between 1.532 and 1.543 at 20°.

Ferrous Fumarate, page 120

Change the *Requirements* entitled *Assay* and *Loss on Drying* to read:

Assay Not less than 97.0% and not more than 101.0% $C_4H_2FeO_4$, calculated on the dried basis.
Loss on Drying Not more than 1.5%.

Change the ninth sentence of the test entitled *Lead, Procedure*, to read:

Combine the acid washes in a small beaker, and adjust the pH with diluted ammonia TS to 2.5 ± 0.2 (measured by means of a glass electrode).

Insert the following before the first sentence of the test entitled *Mercury, Procedure*:

(NOTE: Because mercuric dithizonate is light sensitive, this procedure should be performed in subdued light.)

Fructose, page 130

Change the *Description* to read:

Fructose occurs as white, hygroscopic, odorless, purified crystals or as a purified crystalline powder having a sweet taste. It is a natural constituent of fruit (hence, the name "fruit sugar") and is obtained from glucose in corn syrup by the use of glucose isomerase. Its density is about 1.6. It is soluble in methanol and in ethanol, freely soluble in water, and insoluble in ether.

Replace *Identification Test B* with the following:

The infrared absorption spectrum of a potassium bromide dispersion of the sample, previously dried, exhibits maxima only at the same wavelengths as those of a similar preparation of the USP Fructose Reference Standard.

Change the *Requirements* entitled *Assay* and *Heavy Metals* to read:

Assay Not less than 98.0% and not more than 102.0% Fructose ($C_6H_{12}O_6$), after drying.
Heavy Metals (as Pb) Not more than 5 mg/kg.

Replace the *Requirement* entitled *Hydroxymethylfurfural* with the following:

Hydroxymethylfurfural Not more than 0.1%, calculated on the dried basis.

Insert the following under *Requirements*:

Lead Not more than 0.1 mg/kg.

Replace the last sentence of the *Test* entitled *Assay* to read:

The observed rotation, in degrees (absolute value), multiplied by 1.124 (or 0.562 for the 200-mm tube), represents the weight, in g, of Fructose ($C_6H_{12}O_6$) in the sample taken.

Change the first sentence of the *Test* entitled *Glucose, Glucose Standard Solution*, to read:

Transfer about 300 mg, accurately weighed, of USP Dextrose Reference Standard, previously dried in vacuum at 70° for 4 h, into a 1000-mL volumetric flask, dissolve in and dilute to volume with water, and mix.

Change the third sentence of the *Test* entitled *Glucose, Procedure*, to read:

Allow 30- and 60-s intervals between the addition of *Reagent Solution* to each tube, and add 1.0 mL of the *Reagent Solution* to each of the remaining tubes.

Replace the *Tests* entitled *Hydroxymethylfurfural* and *Heavy Metals* to read:

Hydroxymethylfurfural   Transfer approximately 1 g of the sample, accurately weighed, to a 100-mL volumetric flask, dilute to volume with water, and mix. Read the absorbance of this solution against a water blank at 283 nm in a 1-cm quartz cell in a spectrophotometer. Calculate the percentage of 5-hydroxymethylfurfural (HMF) by the following formula:

$$\% \text{ HMF} = (0.749 \times A)/C,$$

in which A is the absorbance of the sample solution, and C is the concentration, in mg/mL, of the sample solution corrected for ash and moisture.

Heavy Metals   Prepare and test a 4-g sample as directed under the *Heavy Metals Limit Test*, page 512, using 20 µg of lead (Pb) in the control (*Solution A*).

Insert the following under *Tests*:

Lead   Determine as directed under *Method I* in the *Atomic Absorption Spectrophotometric Graphite Furnace Method* under the *Lead Limit Test*, page 168 of THIS SUPPLEMENT, using a 5-g sample.

Garlic Oil, page 132

Change the *Requirements* entitled *Refractive Index* and *Specific Gravity* to read:

Refractive Index   Between 1.550 and 1.580 at 20°.
Specific Gravity   Between 1.050 and 1.095.

Insert the following new monograph to precede the monograph entitled *Geranium Oil, Algerian Type*, page 132:

Gellan Gum

DESCRIPTION

A high molecular weight polysaccharide gum produced by a pure-culture fermentation of a carbohydrate with *Pseudomonas elodea*, and purified by recovery with isopropyl alcohol, dried, and milled. It is a heteropolysaccharide comprising a tetrasaccharide repeating unit of one rhamnose, one glucuronic acid, and two glucose units. The glucuronic acid is neutralized to mixed potassium, sodium, calcium, and magnesium salts. It may contain acyl (glyceryl and acetyl) groups as the *O*-glycosidically linked ester. It occurs as an off-white powder that is soluble in hot or cold deionized water.

REQUIREMENTS

Identification

A. A 1% solution is made by hydrating a 1-g sample in 99 mL of deionized water. The mixture is stirred for about 2 h, using a motorized stirrer and a propeller-type stirring blade. Draw a small amount of the above solution into a wide-bore pipet, and transfer it into a solution of 10% calcium chloride. A tough, worm-like gel will form instantly.
B. To the 1% deionized water solution prepared for *Identification Test A*, add 0.5 g of sodium chloride, heat the solution to 80°, stirring constantly, and hold the temperature at 80° for 1 min. Stop heating and stirring the solution, and allow it to cool to room temperature. A firm gel will form.

Assay   It yields, on the dry basis, not less than 3.3% and not more than 6.8% of carbon dioxide (CO_2).
Arsenic (as As)   Not more than 3 mg/kg.
Ash   Between 4.0% and 12.0%.
Heavy Metals (as Pb)   Not more than 0.003%.
Isopropyl Alcohol   Not more than 0.075%.
Lead   Not more than 2 mg/kg.
Loss on Drying   Not more than 15.0%.

TESTS

Assay   Proceed as directed under *Alginates Assay*, page 463, but use an undried sample of about 1.2 g, accurately weighed.
Arsenic   A *Sample Solution* prepared as directed for organic compounds meets the requirements of the *Arsenic Test*, page 464.
Ash   Accurately weigh about 3 g, previously dried at 105° for 4 h, in a tared crucible, and incinerate it at about 650° until the sample is free from carbon. Cool the crucible and

its contents in a desiccator, weigh, and determine the weight of the ash.

Heavy Metals Prepare and test a 500-mg sample as directed in the *Heavy Metals Test, Method II*, page 512, using a platinum crucible for the ignition and 15 μg of lead ion (Pb) in the control (*Solution A*).

Isopropyl Alcohol Proceed as directed under *Isopropyl Alcohol* in the *Xanthan Gum* monograph, page 347, using a sample of about 5 g, accurately weighed.

Lead A *Sample Solution* prepared with a 1-g sample as directed for organic compounds meets the requirements of the *Lead Limit Test*, page 518, using 2 μg of lead ion (Pb) in the control.

Loss on Drying, page 518 Dry at 105° for 2.5 h.

Functional Use in Foods Stabilizer; thickener.
Packaging and Storage Store in well-closed containers.

Insert the following new monographs to precede the monograph entitled L-*Glutamic Acid*, page 135:

Glucose Syrup

Corn Syrup

DESCRIPTION

A clarified, concentrated, aqueous solution of saccharides obtained by the partial hydrolysis of edible starch by food-grade acids and/or enzymes. Depending on the degree of hydrolysis, it contains varying amounts of D-glucose. When obtained from corn starch, it is commonly designated as corn syrup. It is a sweet, clear white to light yellow viscous liquid miscible in all proportions with water.

REQUIREMENTS

Identification

Add a few drops of a 1 in 20 solution of the sample to 5 mL of hot alkaline cupric tartrate TS. A red precipitate of cuprous oxide is formed.

Assay Not less than 20% reducing sugar content (dextrose equivalent) expressed as D-glucose, calculated on the dried basis.
Arsenic (as As) Not more than 1 mg/kg.
Heavy Metals (as Pb) Not more than 5 mg/kg.
Lead Not more than 0.5 mg/kg.
Residue on Ignition Not more than 0.50%.
Starch Passes test.
Sulfur Dioxide Not more than 40 mg/kg.
Total Solids Not less than 70.0%.

TESTS

Assay Determine as directed in the *Reducing Sugars Assay*, page 169 of THIS SUPPLEMENT.
Arsenic A *Sample Solution* prepared using a 1-g sample meets the requirements of the *Arsenic Test*, page 464, using 1 mL of *Standard Arsenic Solution* in the control (1 μg As).
Heavy Metals A solution of 4 g in 25 mL of water meets the requirements of the *Heavy Metals Test*, page 512, using 20 μg of lead ion (Pb) in the control (*Solution A*).
Lead Determine as directed under *Method I* in the *Atomic Absorption Spectrophotometric Graphite Furnace Method* under the *Lead Limit Test*, page 168 of THIS SUPPLEMENT, using a 5-g sample.
Residue on Ignition, page 533 Ignite 20 g.
Starch To 1 g dissolved in 10 mL of water add 1 drop of iodine TS. A yellow color indicates the absence of soluble starch.
Sulfur Dioxide Determine as directed in the general method, page 170 of THIS SUPPLEMENT, using a 35-g sample.
Total Solids Determine by refractive index as described under *Invert Sugar*, page 84 of the Second Supplement, and using the table on page 172 of THIS SUPPLEMENT.

Functional Use in Foods Nutritive sweetener.
Packaging and Storage Store in tightly closed containers in a dry place.

Glucose Syrup, Dried

Dried Glucose Syrup; Glucose Syrup Solids

DESCRIPTION

Dried Glucose Syrup is a purified, concentrated mixture of nutritive saccharides obtained by the hydrolysis of edible starch and by partially drying the resulting solution (glucose syrup). When obtained from corn starch, it is commonly designated dried corn syrup or corn syrup solids. Depending on the degree of hydrolysis, it contains varying amounts of D-glucose. It is a sweet, white to light yellow powder or granules soluble in water.

REQUIREMENTS

Identification

Add a few drops of a 1 in 20 solution of the sample to 5 mL of hot alkaline cupric tartrate TS. A red precipitate of cuprous oxide is formed.

Assay Not less than 20% reducing sugar content (dextrose equivalent) expressed as D-glucose, calculated on the dried basis.

Arsenic (as As) Not more than 1 mg/kg.
Heavy Metals (as Pb) Not more than 5 mg/kg.
Lead Not more than 0.5 mg/kg.
Residue on Ignition Not more than 0.5%.
Starch Passes test.
Sulfur Dioxide Not more than 40 mg/kg.
Total Solids Not less than 90.0% when the reducing sugar content is 88.0% or greater; not less than 93.0% when the reducing sugar content is between 20.0% and 88.0%.

TESTS

Assay Determine as directed in the *Reducing Sugars Assay*, page 169 of THIS SUPPLEMENT.
Arsenic A solution of 1 g in 35 mL of water meets the requirements of the *Arsenic Test*, page 464, using 1 mL of the *Standard Arsenic Solution* in the control (1 µg As).
Heavy Metals (as Pb) A solution of 2 g in 25 mL of water meets the requirements of the *Heavy Metals Test*, page 512, using 10 µg of lead ion (Pb) in the control (*Solution A*).
Lead Determine as directed under *Method I* in the *Atomic Absorption Spectrophotometric Graphite Furnace Method* under the *Lead Limit Test*, page 168 of THIS SUPPLEMENT, using a 5-g sample.
Residue on Ignition Ignite 1 g as directed in the general method, page 533, using an ignition temperature of 525° for 2 h.
Starch To a 1-g sample dissolved in 10 mL of water add 1 drop of iodine TS. A yellow color indicates the absence of soluble starch.
Sulfur Dioxide Determine as directed in the general method, page 170 of THIS SUPPLEMENT, using a 25-g sample.
Total Solids Determine the water content of an accurately weighed portion as directed under the *Karl Fischer Titrimetric Method*, page 552. Calculate the percent *Total Solids* by the formula:

$$(W_U - W_W)100/W_U,$$

in which W_U is the weight, in mg, of the sample taken; and W_W is the weight, in mg, of water determined.

Functional Use in Foods Nutritive sweetener.
Packaging and Storage Store in tightly closed containers in a dry environment.

L-Glutamic Acid, page 135

Change the *Requirements* entitled *Assay*, *Arsenic*, and *Specific Rotation* to read:

Assay Not less than 98.5% and not more than 101.5% $C_5H_9NO_4$, calculated on the dried basis.
Arsenic (as As) Not more than 1.5 mg/kg.
Specific Rotation $[\alpha]_D^{20°}$: Between +31.5° and +32.2°, calculated on the dried basis.

Delete the *Requirement* entitled *Chloride*.

Replace the *Test* entitled *Assay* with the following:

Assay Dissolve about 200 mg, accurately weighed, in 3 mL of formic acid and 50 mL of glacial acetic acid. Add 2 drops of crystal violet TS, and titrate with 0.1 N perchloric acid to a green endpoint or until the blue color disappears completely. Each mL of 0.1 N perchloric acid is equivalent to 14.71 mg of $C_5H_9NO_4$.

Change the *Tests* entitled *Arsenic* and *Specific Rotation* to read:

Arsenic A *Sample Solution* prepared as directed for organic compounds meets the requirements of the *Arsenic Test*, page 464, using 1.5 mL of the *Standard Arsenic Solution* in the control (1.5 µg As).
Specific Rotation, page 530 $[\alpha]_D^{20°}$: Determine in a solution containing 10 g in sufficient 2 N hydrochloric acid to make 100 mL.

Delete the *Test* entitled *Chloride*.

L-Glutamic Acid Hydrochloride, page 135

Change the *Requirements* entitled *Assay*, *Arsenic*, and *Specific Rotation* to read:

Assay Not less than 98.5% and not more than 101.5% $C_5H_9NO_4 \cdot HCl$, calculated on the dried basis.
Arsenic (as As) Not more than 1.5 mg/kg.
Specific Rotation $[\alpha]_D^{20°}$: Between +25.2° and +25.8°, calculated on the dried basis.

Delete the *Requirement* entitled *Readily Carbonizable Substances*.

Replace the *Test* entitled *Assay* with the following:

Assay Dissolve about 200 mg, accurately weighed, in 3 mL of formic acid and 50 mL of glacial acetic acid. Add 10 mL of mercuric acetate TS and 2 drops of crystal violet TS, and titrate with 0.1 N perchloric acid to the first appearance of a pure green color or until the blue color disappears completely. Each mL of 0.1 N perchloric acid is equivalent to 18.36 mg of $C_5H_9NO_4 \cdot HCl$.

Change the *Tests* entitled *Arsenic* and *Specific Rotation* to read:

Arsenic   A *Sample Solution* prepared as directed for organic compounds meets the requirements of the *Arsenic Test*, page 464, using 1.5 mL of the *Standard Arsenic Solution* in the control (1.5 µg As).

Specific Rotation, page 530   Determine in a solution containing 10 g in sufficient 2 N hydrochloric acid to make 100 mL.

Delete the *Test* entitled *Readily Carbonizable Substances*.

L-Glutamine, page 136

Change the first sentence of the *Identification* to read:

To 5 mL of a 1 in 1000 solution add 1 mL of triketohydrindene hydrate TS, and dissolve by heating in boiling water.

Replace the *Requirements* entitled *Assay* and *Arsenic* to read:

Assay   Not less than 98.5% and not more than 101.5% $C_5H_{10}N_2O_3$, calculated on the dried basis.
Arsenic (as As)   Not more than 1.5 mg/kg.

Change the *Test* entitled *Arsenic* to read:

Arsenic   A *Sample Solution* prepared as directed for organic compounds meets the requirements of the *Arsenic Test*, page 464, using 1.5 mL of the *Standard Arsenic Solution* in the control (1.5 µg As).

Insert the following new monograph to precede the monograph entitled *Glycerin*, page 136:

Glutaraldehyde

Glutaral; 1,5-Pentanedial

$C_5H_8O_2$ Mol wt 100.12

DESCRIPTION

A clear, nearly colorless aqueous solution with a characteristic sharp odor. It is miscible with water. The grades of Glutaraldehyde suitable for food use usually have concentrations between 15% and 50%.

REQUIREMENTS

Identification
  2,4-Dinitrophenylhydrazine Reagent   Add 4 mL of sulfuric acid to 0.8 g of 2,4-dinitrophenylhydrazine, and then add 6 mL of water, dropwise, with swirling. When dissolution is essentially complete, add 20 mL of alcohol, mix, and filter. The filtrate is the reagent.
  Procedure   Add 0.4 mL of the sample to 20 mL of *2,4-Dinitrophenylhydrazine Reagent*. Mix by swirling, and allow the mixture to stand for 5 min. Collect the precipitate on a filter, and rinse thoroughly with alcohol. Dissolve the precipitate in 20 mL of hot ethylene dichloride, filter, and cool the filtrate in an ice bath until crystallization occurs. Collect the precipitate on a filter. Redissolve the precipitate by refluxing with 30 mL of acetone, filter, and cool the filtrate in an ice bath until crystallization occurs. Collect the precipitate on a filter: the 2,4-dinitrophenylhydrazone so obtained melts between 185° and 195° (see *Melting Range*, page 519).

Assay   Not less than 100.0% and not more than 105.0% of the labeled amount of $C_5H_8O_2$.
Arsenic (as As)   Not more than 3 mg/kg.
Heavy Metals (as Pb)   Not more than 10 mg/kg.
pH   Between 3.1 and 4.5.

ADDITIONAL REQUIREMENTS

Labeling   Label to indicate the concentration of Glutaraldehyde.

TESTS

Assay
  Hydroxylamine Hydrochloride Solution   Prepare a 0.5 N solution by dissolving 35.0 g of hydroxylamine hydrochloride in water in a 1-L volumetric flask, dilute to volume with water, and mix.
  Triethanolamine Solution   Prepare a 0.5 N solution by transferring 65 mL (74 g) of 98% triethanolamine to a 1-L volumetric flask, dilute to volume with water, and mix.
  Procedure   Neutralize a volume of *Hydroxylamine Hydrochloride Solution* sufficient for analyzing both the blank and the sample to pH 3.60. Use a suitable autotitrator, and titrate with *Triethanolamine Solution*. (*Caution*: The stirring rate is critical throughout the neutralization and analysis. When stirring is required, ensure adequate mixing without whipping air bubbles into the solution. The stirring speed should be consistent for the sample and blank.)
  Transfer 65.0 mL of the neutralized *Hydroxylamine Hydrochloride Solution* to each of two titration cups. Add a Teflon (or equivalent) stirrer to each cup. Using the autotitrator, add 30.8 mL of the neutralized *Triethanolamine*

Solution to each titration cup, cover, and mix. Using a weighing pipet, introduce into one of the cups a suitable portion of the sample equivalent to about 300 mg of Glutaraldehyde. Mix the solutions thoroughly, and allow the sample and blank to stand at room temperature for at least 60 min but not for more than 90 min.

Titrate the sample and the blank to pH 3.60 with 0.5 N hydrochloric acid, determining the endpoint potentiometrically. Calculate the percentage, by weight, of $C_5H_8O_2$ (w/w) in the sample by the formula:

$$[N(B - A)(0.05006)/W]100,$$

in which B and A are the volumes, in mL, of 0.5 N hydrochloric acid consumed by the blank and sample solutions, respectively; N is the normality of the hydrochloric acid; 0.05006 is the milliequivalent weight, in g per milliequivalent, of Glutaraldehyde; and W is the weight, in g, of sample taken.

Arsenic A *Sample Solution* prepared as directed for organic compounds meets the requirements of the *Arsenic Test*, page 464.

Heavy Metals Prepare and test 2.0 g of the solution as directed under the *Heavy Metals Test*, page 512, using 20 µg of lead ion (Pb) in the control (*Solution A*).

pH, page 531 Determine by the *Potentiometric Method*.

Functional Use in Foods Fixing agent in the immobilization of enzyme preparations; cross-linking agent for microencapsulating flavoring substances.

Packaging and Storage Store in tight, light-resistant containers protected from heat.

Insert the following new monographs to precede the monograph entitled *Glycine*, page 140:

Glyceryl Behenate

Glyceryl Tribehenate; Glyceryl Tridocosanoate

DESCRIPTION

Glyceryl Behenate is a mixture of fatty acid glycerides, primarily glyceryl esters of behenic acid. It is a fine powder that melts at about 70°. It is soluble in chloroform and practically insoluble in water and in alcohol.

REQUIREMENTS

Identification

A. (*Caution*: Ether is highly volatile and flammable. Its vapor, when mixed with air and ignited, may explode.)

Solvent Mixture Prepare a chloroform–acetone (96:4) mixture.

Chromatographic Plates Use suitable thin-layer chromatographic plates (see *Thin-Layer Chromatography*, page 474) coated with a 0.25-mm layer of chromatographic silica gel. Pretreat the plates by placing them in a chromatographic chamber saturated with ether. Remove the plates from the chamber, allow the ether to evaporate, and immerse them in a 2.5% solution of boric acid in alcohol. After about 1 min, withdraw the plates and allow them to dry at ambient temperature. Heat to 110° for 30 min to activate the plates, and then keep them in a desiccator.

Procedure Apply 10 µL of a 6% solution of the sample in chloroform and 10 µL of a 6% solution of USP Glyceryl Behenate Reference Standard in chloroform on one of the chromatographic plates. Develop the chromatogram in the solvent mixture until the solvent front has moved about 12 cm. Remove the plate from the developing chamber and allow the solvent to evaporate. Spray the chromatogram with a 0.02% solution of dichlorofluorescein in alcohol. Examine the spots under short-wavelength ultraviolet light: the R_f values of the spots obtained from the test solution correspond to those obtained from the standard solution.

B. Dissolve about 22 mg of the sample in 1 mL of toluene in a screw-cap vial with a Teflon-lined septum. Add about 0.4 mL of 0.2 N methanolic (*m*-trifluoromethylphenyl)-trimethylammonium hydroxide, attach the cap, and mix. Allow the vial to stand at room temperature for not less than 30 min. Introduce a suitable volume into a gas chromatograph equipped with a flame-ionization detector and a column 1.8 m in length and 4 mm in internal diameter packed with a 10% coating of 50% 3-cyanopropyl–50% phenylmethylsilicone (SP 2300 or equivalent) on a silanized siliceous earth support (Supelcoport or equivalent) maintained at a temperature of about 225°. The retention time of the main peak in the resulting chromatogram corresponds to that of the main peak in a similar preparation of USP Glyceryl Behenate Reference Standard chromatographed concomitantly. The ratio of the response of the main peak to the sum of all the responses is not less than 0.83.

Acid Value Not more than 4.
Arsenic (as As) Not more than 3 mg/kg.
Heavy Metals (as Pb) Not more than 10 mg/kg.
Iodine Value Not more than 3.
1-Monoglycerides Content Not less than 12.0% and not more than 18.0%.
Free Glycerin Not more than 1.0%.
Residue on Ignition Not more than 0.1%.
Saponification Value Not less than 145 and not more than 165.

TESTS

Acid Value In a flask, suspend about 10 g of the sample, accurately weighed, in 50 mL of a 1:1 mixture of alcohol and ether that has been neutralized to phenolphthalein with 0.1 N sodium hydroxide. Connect the flask with a suitable condenser and warm, with frequent shaking, for about 10 min. Add 1 mL of phenolphthalein TS, and titrate with 0.1 N sodium hydroxide until the solution remains faintly pink after shaking for 30 s. Calculate the acid value by the formula:

$$56.1V \times N/W,$$

in which V is the volume, in mL, of the 0.1 N sodium hydroxide solution; N is the normality of the sodium hydroxide solution; and W is the weight, in g, of the sample taken.

Arsenic A *Sample Solution* prepared as directed for organic compounds meets the requirements of the *Arsenic Test*, page 464.

Heavy Metals Proceed as directed under *Method II* of the *Heavy Metals Test*, page 513, using a 2-g sample and 20 μg of lead ion (Pb) in the control solution (*Solution A*).

Iodine Value Proceed as directed under *Iodine Value*, page 505, using the *Wijs Method*.

1-Monoglycerides Content Melt the sample at a temperature not higher than 80°, and mix. Using about 1 g, accurately weighed, proceed as directed in the general method, page 506. (NOTE: If the Glyceryl Behenate titration is less than 0.8 volumes of the blank titration, discard and repeat, using a smaller weight of Glyceryl Behenate.) Calculate the percentage of 1-monoglycerides, as glyceryl monobehenate, by the formula:

$$20.73N(B - S)/W,$$

in which 20.73 is one-twentieth of the molecular weight of glyceryl monobehenate; N is the normality of the sodium thiosulfate; B and S are the volumes, in mL, of 0.1 N sodium thiosulfate consumed by the blank and the Glyceryl Behenate, respectively; and W is the weight, in g, of Glyceryl Behenate taken.

Free Glycerin Proceed as directed in the general method, page 504.

Residue on Ignition Ignite 5 g of the sample as directed in the general method, page 533.

Saponification Value Determine as directed under the general method for *Fats and Related Substances*, page 509.

Functional Use in Foods Formulation aid.

Packaging and Storage Store in tight containers at a temperature not higher than 35°.

Glyceryl Monostearate

Monostearin; 1,2,3-Propanetriol Octadecanoate

$C_{21}H_{42}O_4$

DESCRIPTION

Glyceryl Monostearate occurs as a white, wax-like solid, as flakes, or as beads. It is a mixture of Glyceryl Monostearate and glyceryl monopalmitate. It may contain a suitable antioxidant. It is soluble in hot organic solvents such as acetone, alcohol, benzene, and ether and in mineral or fixed oils. It is dispersible in hot water with the aid of soap or suitable surfactants.

REQUIREMENTS

Assay Not less than 90.0% monoglycerides of saturated fatty acids.
Acid Value Not more than 6.
Arsenic (as As) Not more than 3 mg/kg.
Free Glycerin Not more than 1.2%.
Heavy Metals (as Pb) Not more than 10 mg/kg.
Hydroxyl Value Between 300 and 330.
Iodine Value Not more than 3.
Lead Not more than 1 mg/kg.
Melting Range Not below 65°.
Residue on Ignition Not more than 0.1%.
Saponification Value Not less than 150 and not more than 165.

TESTS

Assay
Propionating Reagent Mix 10 mL of pyridine and 20 mL of propionic anhydride.
Internal Standard Solution Transfer about 400 mg of hexadecyl hexadecanoate, accurately weighed, to a 100-mL volumetric flask, dissolve in chloroform, dilute with chloroform to volume, and mix.
Standard Preparation Transfer about 50 mg of USP Monoglycerides Reference Standard, accurately weighed, to a 25-mL conical flask, add by pipet 5 mL of *Internal Standard Solution*, and mix. When solution is complete, immerse the flask in a water bath maintained at a temperature between 45° and 50°, and volatilize the chloroform with the aid of a stream of nitrogen. Add 3.0 mL of *Propionating Reagent*, and heat the flask on a hot plate at 75° for 30 min. Evaporate the reagents with the aid of a stream of nitrogen and gentle steam heat. Add 15 mL of chloroform, and swirl to dissolve the residue.
Assay Preparation Transfer about 50 mg of the sample, accurately weighed, to a 25-mL conical flask, and proceed

as directed for *Standard Preparation*, beginning with "add by pipet 5 mL of *Internal Standard Solution*, ..."

Chromatographic System   Under typical conditions, the gas chromatograph is equipped with a flame-ionization detector and contains a 2.4-m × 4-mm borosilicate glass column packed with 2% liquid phase 5% phenyl methyl silicone (SE 52 or equivalent) on an 80- to 100-mesh siliceous earth support (Diatoport S or equivalent). The column is maintained isothermally at a temperature between 270° and 280°, the injection port and detector block are maintained at about 310°, and helium is used as the carrier gas at a flow rate of about 70 mL/min.

System Suitability   Chromatograph 6 to 10 injections of the *Standard Preparation* as directed under *Procedure*: the resolution factor, R, between the peaks for the derivatized glyceryl hexadecanoate and glyceryl octadecanoate is not less than 2.0, and the relative standard deviation of the ratio of the peak area of the derivatized glyceryl octadecanoate to that of the hexadecyl hexadecanoate is not more than 2.0%.

Procedure   Inject a suitable portion of the *Standard Preparation* into a suitable gas chromatograph, and record the chromatogram. Measure the areas under the peaks and record the values of the sum of the areas under the derivatized monoglyceride peaks and of the area under the hexadecyl hexadecanoate peak as A_S and A_D, respectively. Calculate the response factor, F, taken by the formula:

$$(A_S/A_D)(W_D/W_S),$$

in which W_D and W_S are the weights, in mg, of hexadecyl hexadecanoate and USP Monoglycerides Reference Standard, respectively, in the *Standard Preparation*. Similarly, inject a suitable portion of the *Assay Preparation* and record the chromatogram. Measure the areas under the peaks, and record the values of the sum of the areas under the derivatized monoglyceride peaks and of the area under the hexadecyl hexadecanoate peak as a_U and a_D, respectively. Calculate the quantity, in mg, of monoglycerides in the amount of Glyceryl Monostearate taken by the formula:

$$(W_D/F)(a_U/a_D).$$

Acid Value   Determine as directed in the general method, page 503.

Arsenic   A *Sample Solution* prepared as directed for organic compounds meets the requirements of the *Arsenic Test*, page 464.

Free Glycerin

Propionating Reagent   Mix 10 mL of pyridine with 20 mL of propionic anhydride.

Internal Standard Solution   Dissolve a suitable quantity of tributyrin, accurately weighed, in chloroform, and dilute quantitatively with chloroform to obtain a solution with a concentration of about 0.2 mg/mL.

Standard Preparation   Transfer about 15 mg of glycerin and about 50 mg of tributyrin, both accurately weighed, to a glass-stoppered, 25-mL conical flask, add 3 mL of *Propionating Reagent*, and heat at 75° for 30 min. Volatilize the reagents with the aid of a stream of nitrogen at room temperature, add about 12 mL of chloroform, and mix. Dilute about 1 mL of this mixture with chloroform to about 20 mL, and mix.

Test Preparation   Transfer about 50 mg of Glyceryl Monostearate, accurately weighed, to a glass-stoppered, 25-mL conical flask, add by pipet 5 mL of *Internal Standard Solution*, and mix to dissolve. Immerse the flask in a water bath maintained at a temperature between 45° and 50°, and volatilize the chloroform with the aid of a stream of nitrogen. Add 3 mL of *Propionating Reagent*, and heat at 75° for 30 min. Volatilize the reagents with the aid of a stream of nitrogen at room temperature, add about 5 mL of chloroform, and mix.

Chromatographic System   Under typical conditions, the gas chromatograph is equipped with a flame-ionization detector and contains a 2.4-m × 4-mm borosilicate glass column packed with 2% liquid phase consisting of a high molecular weight compound of polyethylene glycol and a diepoxide (Carbowax 20 M or equivalent) on an 80- to 100-mesh siliceous earth support (Chromosorb W AW DMCS or equivalent). The column is maintained isothermally at a temperature between 190° and 200°, the injection port and detector block are maintained at about 300° and 310°, respectively, and helium is used as the carrier gas at a flow rate of about 70 mL/min.

System Suitability   Chromatograph 6 to 10 injections of the *Standard Preparation* as directed under *Procedure*: the resolution factor, R, between the peaks for the derivatized glycerin and tributyrin is not less than 4.0, and the relative standard deviation of the ratio of their peak areas is not more than 2.0%.

Procedure   Inject a suitable portion of the *Standard Preparation* into a suitable gas chromatograph, and record the chromatogram. Measure the areas under the peaks, and record the values of the areas under the tripropionin and tributyrin peaks as A_S and A_D, respectively. Calculate the response factor, F, taken by the formula:

$$(A_D/A_S)(W_S/W_D),$$

in which W_S and W_D are the weights, in mg, of glycerin and tributyrin, respectively, in the *Standard Preparation*. Similarly, inject a suitable portion of the *Test Preparation* and record the chromatogram. Measure the areas under the peaks and record the values of the areas under the tripropionin and tributyrin peaks as a_U and a_D, respectively. Calculate the percentage of glycerin by the formula:

$$100F(a_U/a_D)(w_D/w_U),$$

in which w_D is the weight, in mg, of tributyrin in 5 mL of *Internal Standard Solution*, and w_U is the weight, in mg, of Glyceryl Monostearate in the *Test Preparation*.

Heavy Metals Proceed as directed under *Method II* of the *Heavy Metals Test*, page 513, using a 2-g sample and 20 µg of lead ion (Pb) in the control solution (*Solution A*).

Hydroxyl Value, page 504 Proceed as directed for *Method II*.

Iodine Value Proceed as directed under *Iodine Value*, page 505, using the *Hanus Method*.

Lead Prepare and test a 3-g sample as directed under the *Lead Limit Test*, page 518, using 3 µg of lead ion (Pb) in the control.

Melting Range, page 519 Determine as directed in *Procedure for Class I*.

Residue on Ignition Ignite 5 g of the sample as directed in the general method, page 533.

Saponification Value Determine as directed under the general method for *Fats and Related Substances*, page 509.

Functional Use in Foods Emulsifier.
Packaging and Storage Store in tight, light-resistant containers.

Glycine, page 140

Change the *Requirements* entitled *Assay* and *Arsenic* to read:

Assay Not less than 98.5% and not more than 101.5% $C_2H_5NO_2$, calculated on the dried basis.
Arsenic (as As) Not more than 1.5 mg/kg.

Delete the *Requirement* entitled *Readily Carbonizable Substances*.

Change the *Test* entitled *Arsenic* to read:

Arsenic A *Sample Solution* prepared as directed for organic compounds meets the requirements of the *Arsenic Test*, page 464, using 1.5 mL of the *Standard Arsenic Solution* in the control (1.5 µg As).

Delete the *Test* entitled *Readily Carbonizable Substances*.

Insert the following new monograph to precede the monograph entitled *Gum Guaiac*, page 141:

Gum Ghatti

Indian Gum

DESCRIPTION

Gum Ghatti is the dried gummy exudate from the stems of *Anogeissus latifolia* Wall of the family *Combretaceae*. It is a complex, water-soluble polysaccharide composed of the calcium and magnesium salts of L-arabinose, D-galactose, D-mannose, D-xylose, and D-glucuronic acids in the approximate molar ratio of 10:6:2:1:2. It is light to dark tan and is insoluble in 90% alcohol.

REQUIREMENTS

Identification

Lead Acetate Solution (*Caution*: Use gloves and goggles to avoid contact with skin and eyes. Use an effective fume-removal device or other respiratory protection.) Activate 5 to 60 g of lead (II) oxide by heating it for 2.5 to 3 h in a furnace at 650° to 670° (cooled product should have a lemon color). In a 500-mL Erlenmeyer flask provided with a reflux condenser, boil 80 g of lead acetate trihydrate and 40 g of freshly activated lead (II) oxide with 250 g of water for 45 min. Cool, filter off any residue, and dilute with recently boiled water to a density of 1.25 at 20°. Add 4 mL of water to 1 mL of the lead acetate solution, and filter.

Procedure Add 0.2 mL of the *Lead Acetate Standard Solution* to 5 mL of a cold 1 in 100 aqueous solution of the gum. A slight precipitate or clear solution results that yields an opaque flocculent precipitate upon the addition of 1 mL of 3 *N* ammonium hydroxide.

Arsenic (as As) Not more than 3 mg/kg.
Ash (Acid-Insoluble) Not more than 1.75%.
Ash (Total) Not more than 6.0%.
Heavy Metals (as Pb) Not more than 40 mg/kg.
Insoluble Matter Not more than 1.0%.
Lead Not more than 10 mg/kg.
Loss on Drying Not more than 14.0%.
Viscosity A 5% solution exhibits a viscosity measured in centipoises within the range stated on the label.

ADDITIONAL REQUIREMENTS

Labeling Label to indicate the viscosity, in centipoises.

TESTS

Arsenic A *Sample Solution* prepared as directed for organic compounds meets the requirements of the *Arsenic Test*, page 464.

Ash (Acid-Insoluble) Determine as directed in the general method, page 466.

Ash (Total) Determine as directed in the general method, page 466.

Heavy Metals Prepare and test a 500-mg sample as directed in *Method II* under the *Heavy Metals Test*, page 513, using 20 μg of lead ion (Pb) in the control (*Solution A*).

Insoluble Matter Dissolve a 5-g sample, accurately weighed, in about 100 mL of water contained in a 250-mL Erlenmeyer flask, add 10 mL of hydrochloric acid TS, and boil gently for 15 min. Filter the hot solution, using suction through a tared filtering crucible, wash thoroughly with hot water, dry at 105° for 2 h, and weigh.

Lead A *Sample Solution* prepared as directed for organic compounds meets the requirements of the *Lead Limit Test*, page 518, using 10 μg of lead ion (Pb) in the control.

Loss on Drying, page 518 Dry at 105° for 5 h. After drying, unground samples should be powdered to pass through a No. 40 sieve and mixed well before weighing.

Viscosity Determine as directed under *Viscosity of Sodium Carboxymethylcellulose*, page 550, at 75°, using spindle No. 2 at 60 rpm.

Functional Use in Foods Emulsifier and emulsifier salt.
Packaging and Storage Store in well-closed containers.

Insert the following new monograph to precede the monograph entitled *Heptylparaben*, page 142:

Helium

He   Mol wt 4.0026

DESCRIPTION

A colorless, odorless gas that is not combustible and does not support combustion. Very slightly soluble in water. One L of the gas weighs about 180 mg at 0° and at a pressure of 760 mm of mercury.

REQUIREMENTS

NOTE: Reduce the container pressure by means of a regulator. Measure the gas with a gas volume meter downstream from the pertinent detector tube to minimize contamination of or change to the specimens.

Identification

The flame of a burning splinter of wood is extinguished when inserted into an inverted test tube filled with Helium. (NOTE: Use caution.) A small balloon filled with Helium shows buoyancy.

Assay Not less than 99.0% He, by volume.
Air Not more than 1.0%, by volume.
Carbon Monoxide Not more than 10 ppm, by volume.
Odor Passes test.

TESTS

Assay Introduce a specimen of Helium into a gas chromatograph by means of a gas sampling valve. Select the operating conditions of the gas chromatograph so that the standard peak signal resulting from the following procedure corresponds to not less than 70% of the full-scale reading. Preferably use an apparatus corresponding to the general type in which the column is 6 m in length and 4 mm in inside diameter and is packed with porous polymer beads (PoraPak Q or equivalent), which permit complete separation of nitrogen and oxygen from Helium, although nitrogen and oxygen may not be separated from each other. Use industrial-grade Helium (99.99%) as the carrier gas, with a thermal-conductivity detector, and control the column temperature at 60°. The peak response produced by the assay specimen exhibits a retention time corresponding to that produced by an air–Helium certified standard (a mixture of 1.0% air in industrial-grade helium is available from most suppliers) and indicates not more than 1.0% air, by volume, when compared with the peak response of the air–Helium certified standard, and not less than 99.0% He, by volume.

Air Determine as directed in the *Assay*.

Carbon Monoxide Pass 1050 ± 50 mL through a carbon monoxide detector tube (described on page 173 of THIS SUPPLEMENT) at the rate specified for the tube. The indicator change corresponds to not more than 10 ppm, by volume.

Odor Carefully open the container valve to produce a moderate flow of gas. Do not direct the gas stream toward the face, but deflect a portion of the stream toward the nose. No appreciable odor is discernible.

Functional Use in Foods Processing aid.
Packaging and Storage Store in appropriate gas cylinders.

High-Fructose Corn Syrup, page 51, Second Supplement

Change the *Test* entitled *Heavy Metals* to read:

Heavy Metals (as Pb) Not more than 5 mg/kg.

Insert the following under *Requirements*:

Lead Not more than 0.5 mg/kg.

Change the name of the *Requirement* entitled *Solids* to read:

Total Solids

Change the first sentence of the *Test* entitled *Assay, Standardization*, to read:

Prepare a standard solution containing a total of about 10% solids using sugars of known purity (e.g., USP Fructose Reference Standard; USP Dextrose Reference Standard or NIST Standard Reference Material; maltose, Aldrich Chemical Company; or any equivalent) that approximates, on the dry basis, the composition of the sample to be analyzed.

Change the *Test* entitled *Heavy Metals* to read:

Heavy Metals   Prepare and test a 4-g sample as directed in *Method II* under the *Heavy Metals Test*, page 513, using 20 µg of lead ion (Pb) in the control (*Solution A*) and 500° as the ignition temperature.

Replace the *Test* entitled *Lead* with the following:

Lead   Determine as directed in *Method I* in the *Atomic Absorption Spectrophotometric Graphite Furnace Method* under the *Lead Limit Test*, page 168 of THIS SUPPLEMENT, using a 5-g sample.

Change the name of the *Test* entitled *Solids* to read:

Total Solids

Change the *Tests* entitled *Residue on Ignition*, *Sulfur Dioxide*, and *Total Solids* to read:

Residue on Ignition   Ignite 10 g as directed under *Residue on Ignition*, page 533.
Sulfur Dioxide   Determine as directed in the general method, page 170 of THIS SUPPLEMENT, using a 50-g sample.
Total Solids   Determine by refractive index using the table, page 172 in THIS SUPPLEMENT.

L-Histidine, page 143

Change the *Requirements* entitled *Assay*, *Arsenic*, and *Specific Rotation* to read:

Assay   Not less than 98.5% and not more than 101.5% $C_6H_9N_3O_2$, calculated on then dried basis.
Arsenic (as As)   Not more than 1.5 mg/kg.
Specific Rotation $[\alpha]_D^{20°}$: Between +11.5° and +13.5°, calculated on the dried basis.

Change the *Tests* entitled *Assay*, first sentence, and *Arsenic* to read:

Assay   Dissolve about 150 mg of the sample, previously dried at 105° for 3 h and accurately weighed, in 3 mL of formic acid and 50 mL of glacial acetic acid, and titrate with 0.1 *N* perchloric acid, determining the endpoint potentiometrically.
Arsenic   A *Sample Solution* prepared as directed for organic compounds meets the requirements of the *Arsenic Test*, page 464, using 1.5 mL of the *Standard Arsenic Solution* in the control (1.5 µg As).

L-Histidine Monohydrochloride, page 143

Change the *Requirements* entitled *Assay* and *Arsenic* to read:

Assay   Not less than 98.5% and not more than 101.5% $C_6H_9N_3O_2 \cdot HCl \cdot H_2O$, calculated on the dried basis.
Arsenic (as As)   Not more than 1.5 mg/kg.

Change the *Test* entitled *Arsenic* to read:

Arsenic   A *Sample Solution* prepared as directed for organic compounds meets the requirements of the *Arsenic Test*, page 464, using 1.5 mL of the *Standard Arsenic Solution* in the control (1.5 µg As).

Invert Sugar, Second Supplement, page 53

Change the *Requirements* entitled *Assay*, *Arsenic*, *Heavy Metals*, and *Lead* to read:

Assay   Percent sucrose and Invert Sugar content shall be as labeled by the manufacturer.
Arsenic (as As)   Not more than 1 mg/kg.
Heavy Metals (as Pb)   Not more than 5 mg/kg.
Lead   Not more than 0.5 mg/kg.

Insert the following heading after the *Requirements* section:

ADDITIONAL REQUIREMENTS

Insert the following *Additional Requirement* for *Labeling*:

Labeling   Label to indicate the percent sucrose and Invert Sugar.

Change the *Test* entitled *Arsenic* to read:

Arsenic   A *Sample Solution* prepared as directed using a 1-g sample meets the requirements of the *Arsenic Test*, page 464, using 1 mL of the *Standard Arsenic Solution* in the control (1 µg As).

Replace the *Tests* for *Heavy Metals* and *Lead* as follows:

Heavy Metals   A solution of 4 g in 25 mL of water meets the requirements of the *Heavy Metals Test*, page 512, using 20 µg of lead ion (Pb) in the control (*Solution A*).

Lead   Determine as directed under *Method I* in the *Atomic Absorption Spectrophotometric Graphite Furnace Method* under the *Lead Limit Test*, page 168 of THIS SUPPLEMENT, using a 5-g sample.

Insert the following new monograph to precede the monograph entitled *Isobutylene-Isoprene Copolymer*, page 153:

Isobutane

CH$_3$CH(CH$_3$)$_2$ Mol wt 58.12

C$_4$H$_{10}$

DESCRIPTION

A colorless, odorless, flammable gas (boiling temperature is about –11°). The vapor pressure at 21° is approximately 2950 mm of mercury (31 psi).

REQUIREMENTS

Caution: Isobutane is highly flammable and explosive. Observe precautions and perform sampling and analytical operations in a well-ventilated fume hood.

Identification

A. The infrared absorption spectrum exhibits maxima, among others, at about the following wavelengths, in µm: 3.4 (vs), 6.8 (s), 7.2 (m), and 10.9 (m).
B. The vapor pressure of a test specimen, obtained as directed in the *Sampling Procedure* and determined at 21° by means of a suitable pressure gauge, is between 303 and 331 kPa absolute (44 and 48 psia, respectively).

Assay   Not less than 95.0% C$_4$H$_{10}$.
Acidity of Residue   Passes test.
High-Boiling Residue   Not more than 5 mg/kg.
Sulfur Compounds   Passes test.
Water   Not more than 10 mg/kg.

TESTS

Sampling Procedure, Assay, Acidity of Residue, High-Boiling Residue, Sulfur Compounds, and **Water**   Proceed as directed for these tests under *Butane*, page 101 of THIS SUPPLEMENT, except substitute Isobutane wherever Butane is specified in the text.

Functional Use in Foods   Propellant; aerating agent.
Packaging and Storage   Store in tight cylinders protected from excessive heat.

DL-Isoleucine, page 154

Insert the following at the end of the *Description*:

It is optically inactive.

Change the *Requirements* entitled *Assay* and *Arsenic* to read:

Assay   Not less than 98.5% and not more than 101.5% C$_6$H$_{13}$NO$_2$, calculated on the dried basis.
Arsenic (as As)   Not more than 1.5 mg/kg.

Change the *Test* entitled *Arsenic* to read:

Arsenic   A *Sample Solution* prepared as directed for organic compounds meets the requirements of the *Arsenic Test*, page 464, using 1.5 mL of the *Standard Arsenic Solution* in the control (1.5 µg As).

L-Isoleucine, page 154

Change the *Requirements* entitled *Assay*, *Arsenic*, and *Specific Rotation* to read:

Assay   Not less than 98.5% and not more than 101.5% C$_6$H$_{13}$NO$_2$, calculated on the dried basis.
Arsenic (as As)   Not more than 1.5 mg/kg.
Specific Rotation $[\alpha]_D^{20°}$: Between +38.6° and +41.5°, calculated on the dried basis.

Change the *Test* entitled *Arsenic* to read:

Arsenic   A *Sample Solution* prepared as directed for organic compounds meets the requirements of the *Arsenic Test*, page 464, using 1.5 mL of the *Standard Arsenic Solution* in the control (1.5 µg As).

Insert the following new monograph to precede the monograph entitled *Labdanum Oil*, page 158:

Konjac Flour

Konjac; Konnyaku; Konjac Gum

DESCRIPTION

A hydrocolloidal polysaccharide obtained from the tubers of various species of *Amorphophallus*. Konjac Flour is a high molecular weight, nonionic glucomannan primarily consisting of mannose and glucose at a respective molar ratio of approximately 1.6:1.0. It is a slightly branched polysaccharide connected by β-1,4 linkages and has an average molecular weight of 200,000 to 2,000,000 daltons. Acetyl groups along the glucomannan backbone contribute to solubility properties and are located, on average, every 9 to 19 sugar units. The typical powder is cream to light tan in color.

Konjac Flour is dispersible in hot or cold water and forms a highly viscous solution with a pH between 4.0 and 7.0. Solubility is increased by heat and mechanical agitation. Addition of mild alkali to the solution results in the formation of a heat-stable gel that resists melting, even under extended heating conditions.

REQUIREMENTS

Identification

A. *Microscopic Test*   Stain about 0.1 g of the sample with 0.01% methylene blue powder in 50% isopropyl alcohol, and observe microscopically. Konjac Flour may be identified by the presence of flattened elliptical particles, which are generally 100 to 500 μm in length along the long axis. Unground Konjac Flour is clearly distinguished from other hydrocolloids by the presence of sac-like cells that contain glucomannan. The surface of these cells has a reticulated structure. Particles of Konjac Flour are also birefringent under polarized light. These visual characteristics may remain even if the sample is finely ground, but are less pronounced.

B. *Gel Test*   At room temperature, add 5 mL of a 4% sodium borate solution to a 1% solution of the sample in a test tube, and shake vigorously. If Konjac Flour is present, a gel forms. (Konjac Flour solutions gel in the presence of sodium borate, similar in reaction to that of other galactomannans such as guar gum and locust bean gum.) The *Heat-Stable Gel Test*, below, distinguishes Konjac Flour from guar and locust bean gums.

C. *Heat-Stable Gel Test*   Prepare a 2% solution of the sample by heating it in a boiling water bath for 30 min with continuous agitation and then cooling the solution to room temperature. For each g of the sample used to prepare the 2% solution, add 1 mL of 10% potassium carbonate solution to the fully hydrated sample at ambient temperature. Heat the mixture in a water bath to 85°, and hold quiescently for 2 h without agitation. Konjac Flour forms a thermally stable gel under these conditions. Related hydrocolloids such as guar gum and locust bean gum do not form thermally stable gels and are negative by this test.

Arsenic (as As)   Not more than 3 mg/kg.
Ash (Total)   Not more than 5.0%.
Carbohydrate (Total)   Not less than 75.0%.
Heavy Metals   Not more than 10 mg/kg.
Lead   Not more than 5 mg/kg.
Loss on Drying   Not more than 15.0%.
Protein   Not more than 8.0%.

TESTS

Arsenic   A *Sample Solution* prepared as directed for organic compounds meets the requirements of the *Arsenic Test*, page 464.
Ash (Total)   Determine as directed in the general method, page 466.
Carbohydrate (Total)   The remainder, after subtracting from 100% the sum of the percentages of *Ash*, *Loss on Drying*, and *Protein*, represents the percentage of carbohydrates (glucomannans) in the sample.
Heavy Metals   Prepare and test a 2-g sample as directed in *Method II* under the *Heavy Metals Test*, page 513, using 20 μg of lead ion (Pb) in the control (*Solution A*).
Lead   A 2-g sample prepared as directed for organic compounds meets the requirements of the *Lead Limit Test*, page 518, using 10 μg of lead ion (Pb) in the control.
Loss on Drying, page 518   Dry at 105° for 5 h.
Protein   Transfer about 3.5 g, accurately weighed, into a 500-mL Kjeldahl flask, and proceed as directed under *Nitrogen Determination*, page 521. Percent protein equals percent $N \times 5.7$.

Packaging and Storage   Store cool and dry in a closed container away from direct heat and sunlight.
Functional Use in Foods   Gelling agent; thickener; film former; emulsifier; stabilizer.

Lactated Mono-Diglycerides, page 159

Replace the monograph title with the following:

Glyceryl–Lacto Esters of Fatty Acids

Insert the following *Synonym*:

Lactated Mono-Diglycerides

Change the *Tests* entitled *Arsenic* and *Heavy Metals* to read:

Arsenic (as As)   Not more than 1 mg/kg.
Heavy Metals (as Pb)   Not more than 5 mg/kg.

Insert the following under *Requirements*:

Lead   Not more than 0.5 mg/kg.
Residue on Ignition   Not more than 0.1%.
Unsaponifiable Matter   Not more than 2%.

Change the specifications conforming to the representations of the vendor to read:

Acid Value, Free Glycerin, 1-Monoglyceride Content, Total Lactic Acid, and **Water**.

Change the *Tests* entitled *Arsenic* and *Heavy Metals* to read:

Arsenic   A *Sample Solution* prepared as directed for organic compounds meets the requirements of the *Arsenic Test*, page 464, using 1 mL of the *Standard Arsenic Solution* (1 µg As).
Heavy Metals   Prepare and test a 4-g sample as directed in *Method II* under the *Heavy Metals Test*, page 513, using 20 µg of lead ion (Pb) in the control (*Solution A*).

Insert the following under *Tests*:

Lead   Prepare and test a 4-g sample as directed under the *Lead Limit Test*, page 518, using 2 µg of lead ion (Pb) in the control.
Residue on Ignition, page 533   Ignite 1 g as directed in the general method.
Unsaponifiable Matter, page 509   Determine as directed in the general method.

Lactic Acid, page 159

Change the *Synonym* to read:

α-Hydroxypropionic Acid; 2-Hydroxypropionic Acid

Insert the following *Formula* and *Molecular Weight*:

$C_3H_6O_3$ Mol wt 90.08

Change the third sentence of the *Description* to read:

It is usually available in solutions containing the equivalent of from 50% to 90% Lactic Acid.

Change the *Requirement* entitled *Arsenic* to read:

Arsenic (as As)   Not more than 1 mg/kg.

Replace the *Requirement* entitled *Cyanide* as follows:

Cyanide   Not more than 5 mg/kg.

Insert the following *Requirement* for *Lead*:

Lead   Not more than 5 mg/kg.

Change the second sentence of the *Test* entitled *Cyanide, Cyanide Standard Solution*, to read:

Transfer a 10-mL aliquot into a 1000-mL volumetric flask, dilute to volume with 0.1 N sodium hydroxide, and mix.

Change the last sentence of the *Test* entitled *Cyanide, Procedure*, to read:

The absorbance of the *Sample Solution*, determined at 480 nm with a suitable spectrophotometer, is no greater than that of the *Cyanide Standard Solution*.

Change the *Test* entitled *Arsenic* to read:

Arsenic   A *Sample Solution* prepared as directed for organic compounds meets the requirements of the *Arsenic Test*, page 464, using 1 mL of the *Standard Arsenic Solution* in the control (1 µg As).

Insert the following under *Tests*:

Lead   A 4-g sample prepared as directed for organic compounds meets the requirements of the *Lead Limit Test*, page 518, using 20 µg of lead ion (Pb) in the control.

DL-Leucine, page 170

Insert the following at the end of the *Description*:

It is optically inactive.

Change the *Requirements* entitled *Assay* and *Arsenic* to read:

Assay   Not less than 98.5% and not more than 101.5% $C_6N_{13}NO_2$, calculated on the dried basis.
Arsenic (as As)   Not more than 1.5 mg/kg.

Change the *Test* entitled *Arsenic* to read:

Arsenic   A *Sample Solution* prepared as directed for organic compounds meets the requirements of the *Arsenic Test*, page 464, using 1.5 mL of the *Standard Arsenic Solution* in the control (1.5 µg As).

L-Leucine, page 171

Change the *Requirements* entitled *Assay, Arsenic, Heavy Metals,* and *Specific Rotation* to read:

Assay Not less than 98.5% and not more than 101.5% $C_6H_{13}NO_2$, calculated on the dried basis.
Arsenic (as As) Not more than 1.5 mg/kg.
Heavy Metals (as Pb) Not more than 20 mg/kg.
Specific Rotation $[\alpha]_D^{20°}$: Between +14.5° and +16.5°, calculated on the dried basis.

Delete the *Requirements* entitled *Iron, Methionine, Nitrogen,* and *Tyrosine*.

Change the second sentence of the *Test* entitled *Assay* to read:

Dissolve the sample in 3 mL of formic acid and about 50 mL of glacial acetic acid, add 2 drops of crystal violet TS, and titrate with 0.1 N perchloric acid to a bluish green endpoint.

Change the *Tests* entitled *Arsenic, Heavy Metals,* and *Specific Rotation* to read:

Arsenic A *Sample Solution* prepared as directed for organic compounds meets the requirements of the *Arsenic Test,* page 464, using 1.5 mL of the *Standard Arsenic Solution* in the control (1.5 µg As).
Heavy Metals Prepare and test a 1-g sample as directed in *Method II* under the *Heavy Metals Test,* page 513, using 20 µg of lead ion (Pb) in the control (*Solution A*).
Specific Rotation, page 530 Determine in a solution containing 4 g in sufficient 6 N hydrochloric acid to make 100 mL.

Delete the *Tests* entitled *Iron, Methionine, Nitrogen,* and *Tyrosine*.

L-Lysine Monohydrochloride, page 176

Change the *Requirements* entitled *Assay, Arsenic,* and *Specific Rotation* to read:

Assay Not less than 98.5% and not more than 101.5% $C_6H_{14}N_2O_2 \cdot HCl$, calculated on the dried basis.
Arsenic (as As) Not more than 1.5 mg/kg.
Specific Rotation $[\alpha]_D^{20°}$: Between +20.3° and +21.3°, calculated on the dried basis.

Change the *Test* entitled *Arsenic* to read:

Arsenic A *Sample Solution* prepared as directed for organic compounds meets the requirements of the *Arsenic Test,* page 464, using 1.5 mL of the *Standard Arsenic Solution* in the control (1.5 µg As).

Magnesium Chloride, page 177

Change the *Formula* and the *Molecular Weight* to read:

$MgCl_2 \cdot xH_2O$ Mol wt (anhydrous) 95.21

Replace the *Description* with the following:

Magnesium chloride contains two or six molecules of water of hydration. The dihydrate is in the form of small, white, hygroscopic granules. The hexahydrate is in the form of colorless, deliquescent flakes or crystals. Both forms are very soluble in water and freely soluble in alcohol.

Replace the *Requirement* entitled *Assay* with the following:

Assay *Dihydrate*: not less than 95.0% and not more than 100.5% $MgCl_2 \cdot 2H_2O$; *hexahydrate*: not less than 99.0% and not more than 105.0% $MgCl_2 \cdot 6H_2O$.

Change the *Requirement* entitled *Sulfate* to read:

Sulfate Not more than 0.03%.

Insert the following heading after the *Requirements* section:

ADDITIONAL REQUIREMENTS

Insert the following under *Additional Requirements*:

Labeling Label to indicate whether it is the dihydrate or the hexahydrate.

Change the *Tests* entitled *Assay* and *Sulfate* to read:

Assay Dissolve about 300 mg of the dihydrate, or about 450 mg of the hexahydrate, accurately weighed, in 25 mL of water, add 5 mL of ammonia–ammonium chloride buffer TS and 0.1 mL of eriochrome black TS, and titrate with 0.05 M disodium EDTA until the solution turns blue. Each mL of 0.05 M disodium EDTA is equivalent to 6.562 mg of $MgCl_2 \cdot 2H_2O$ or 10.16 mg of $MgCl_2 \cdot 6H_2O$.
Sulfate, page 471 Any turbidity produced by a 1-g sample does not exceed that shown in a control containing 300 µg of sulfate (SO_4).

Insert the following new monograph to precede the monograph entitled *Maltol*, page 184:

Maltodextrin

DESCRIPTION

Maltodextrin is a purified, concentrated, nonsweet, nutritive mixture of saccharide polymers obtained by the partial hydrolysis of edible starch. It occurs as a white, slightly hygroscopic powder, or as granules of similar description, or as a clear to hazy solution in water. Powders or granules are freely soluble or readily dispersible in water.

REQUIREMENTS

Identification

To 5 mL of hot alkaline cupric tartrate TS add a few drops of a 1 in 10 solution of the sample. A red precipitate of cuprous oxide forms.

Assay Not more than 20% reducing sugar content (dextrose equivalent) expressed as D-glucose, calculated on the dried basis.
Arsenic (as As) Not more than 1 mg/kg.
Heavy Metals (as Pb) Not more than 5 mg/kg.
Lead Not more than 0.5 mg/kg.
Protein (Total) Not more than 0.1%.
Residue on Ignition Not more than 0.5%.
Sulfur Dioxide Not more than 40 mg/kg.
Total Solids
  Powders and granules Not less than 94%.
  Liquids Not less than 65%.

TESTS

Assay Determine as directed in the *Reducing Sugars Assay*, page 169 of THIS SUPPLEMENT.
Arsenic A *Sample Solution* prepared using a 1-g sample meets the requirements of the *Arsenic Test*, page 464, using 1 mL of the *Standard Arsenic Solution* in the control (1 μg of As).
Heavy Metals A solution of 2 g in 25 mL of water meets the requirements of the *Heavy Metals Test*, page 512, using 10 μg of lead ion (Pb) in the control (*Solution A*).
Lead Determine as directed under *Method I* in the *Atomic Absorption Spectrophotometric Graphite Furnace Method* under the *Lead Limit Test*, page 168 of THIS SUPPLEMENT, using a 5-g sample.
Protein (Total) Determine as directed under *Nitrogen Determination* (*Kjeldahl Method*), page 521. Protein content is $N \times 6.25$.
Residue on Ignition, page 533 Ignite 1 g as directed in the general method.
Sulfur Dioxide Determine as directed in the general method, page 170 of THIS SUPPLEMENT.
Total Solids Proceed as directed under *Total Solids*, page 84 of the Second Supplement.

Functional Use in Foods Anticaking and free-flowing agent; formulation aid; processing aid; bulking agent; stabilizer and thickener; surface-finishing agent.
Packaging and Storage Keep dry, and store at ambient temperatures.

DL-Methionine, page 193

Change the *Requirements* entitled *Assay* and *Arsenic* to read:

Assay Not less than 98.5% and not more than 101.5% $C_5H_{11}NO_2S$, calculated on the dried basis.
Arsenic (as As) Not more than 1.5 mg/kg.

Change the *Test* entitled *Arsenic* to read:

Arsenic A *Sample Solution* prepared as directed for organic compounds meets the requirements of the *Arsenic Test*, page 464, using 1.5 mL of the *Standard Arsenic Solution* in the control (1.5 μg As).

L-Methionine, page 193

Change the *Requirements* entitled *Assay*, *Arsenic*, and *Specific Rotation* to read:

Assay Not less than 98.5% and not more than 101.5% $C_5H_{11}NO_2S$, calculated on the dried basis.
Arsenic (as As) Not more than 1.5 mg/kg.
Specific Rotation $[\alpha]_D^{20°}$: Between +21.0° and +25.0°, calculated on the dried basis.

Change the *Tests* entitled *Arsenic* and *Specific Rotation* to read:

Arsenic A *Sample Solution* prepared as directed for organic compounds meets the requirements of the *Arsenic Test*, page 464, using 1.5 mL of the *Standard Arsenic Solution* in the control (1.5 μg As).
Specific Rotation, page 530 Determine in a solution containing 2 g in sufficient 6 N hydrochloric acid to make 100 mL.

Monoammonium L-Glutamate, page 200

Insert the following at the end of the *Description*:

The pH of a 1 in 20 solution is between 6.0 and 7.0.

Change the *Requirements* entitled *Assay* and *Arsenic* to read:

Assay Not less than 98.5% and not more than 101.5% $C_5H_{12}N_2O_4 \cdot H_2O$.
Arsenic (as As) Not more than 1.5 mg/kg.

Delete the *Requirement* entitled *pH of a 1 in 20 Solution*.

Replace the *Requirement* entitled *Specific Rotation* with the following:

Specific Rotation $[\alpha]_D^{20°}$: Not less than +25.4° and not more than +26.4°.

Change the *Tests* entitled *Arsenic* and *Specific Rotation* to read:

Arsenic A *Sample Solution* prepared as directed for organic compounds meets the requirements of the *Arsenic Test*, page 464, using 1.5 mL of the *Standard Arsenic Solution* in the control (1.5 μg As).
Specific Rotation, page 530 Determine in a solution containing 10 g in sufficient 2 N hydrochloric acid to make 100 mL.

Delete the *Test* entitled *pH of a 1 in 20 Solution*.

Monopotassium L-Glutamate, page 202

Insert the following at the end of the *Description*:

The pH of a 1 in 50 solution is between 6.7 and 7.3.

Change the *Requirements* entitled *Assay*, *Arsenic*, and *Specific Rotation* to read:

Assay Not less than 98.5% and not more than 101.5% $C_5H_8KNO_4 \cdot H_2O$.
Arsenic (as As) Not more than 1.5 mg/kg.
Specific Rotation $[\alpha]_D^{20°}$: Not less than +22.5° and not more than +24.0°.

Delete the *Requirements* entitled *Chloride* and *pH of a 1 in 50 Solution*.

Change the *Tests* entitled *Arsenic* and *Specific Rotation* to read:

Arsenic A *Sample Solution* prepared as directed for organic compounds meets the requirements of the *Arsenic Test*, page 464, using 1.5 mL of the *Standard Arsenic Solution* in the control (1.5 μg As).
Specific Rotation, page 530 Determine in a solution containing 10 g in sufficient 2 N hydrochloric acid to make 100 mL.

Delete the *Tests* entitled *Chloride* and *pH of a 1 in 50 Solution*.

Monosodium L-Glutamate, page 203

Add the following to the end of the *Description*:

The pH of a 1 in 20 solution is between 6.7 and 7.2.

Change the *Requirements* entitled *Assay*, *Arsenic*, and *Specific Rotation* to read:

Assay Not less than 98.5% and not more than 101.5% $C_5H_8NNaO_4 \cdot H_2O$.
Arsenic (as As) Not more than 1.5 mg/kg.
Specific Rotation $[\alpha]_D^{20°}$: Not less than +24.8° and not more than +25.3°, or $[\alpha]_D^{25°}$: Not less than +29.7° and not more than +30.2°.

Delete the *Requirement* entitled *pH of a 1 in 20 Solution*.

Change the *Tests* entitled *Arsenic* and *Specific Rotation* to read:

Arsenic A *Sample Solution* prepared as directed for organic compounds meets the requirements of the *Arsenic Test*, page 464, using 1.5 mL of the *Standard Arsenic Solution* in the control (1.5 μg As).
Specific Rotation Determine in a solution containing 10 g in sufficient 2 N hydrochloric acid to make 100 mL.

Delete the *Test* entitled *pH of a 1 in 20 Solution*.

Insert the following new monograph to precede the monograph entitled *Niacin*, page 205:

Natamycin

Pimaricin

$C_{33}H_{47}NO_{13}$ Mol wt 665.73

DESCRIPTION

Off-white to cream-colored powder, which may contain up to 3 moles of water. It melts with decomposition at about 280°. Practically insoluble in water, slightly soluble in methanol, and soluble in glacial acetic acid and in dimethylformamide.

REQUIREMENTS

Identification

Transfer 50 mg, accurately weighed, to a 200-mL volumetric flask, add 5.0 mL of water, and moisten the specimen. Add 100 mL of a 1 in 1000 solution of glacial acetic acid in methanol, and shake by mechanical means in the dark until dissolved. Dilute with the acetic acid–methanol solution to volume, and mix. Transfer 2.0 mL of this solution to a 100-mL volumetric flask, dilute with the acetic acid–methanol solution to volume, and mix: the ultraviolet absorption spectrum of the solution so obtained exhibits maxima and minima at the same wavelengths as those of a similar solution of USP Natamycin Reference Standard, concomitantly measured.

Assay Not less than 97.0% and not more than 102.0% $C_{33}H_{47}NO_{13}$, calculated on the anhydrous basis.
Arsenic (as As) Not more than 1 mg/kg.
Heavy Metals (as Pb) Not more than 20 mg/kg.
pH Between 5.0 and 7.5.
Specific Rotation $[\alpha]_D^{20°}$: Between +276° and +280°.
Water Between 6.0% and 9.0%.

TESTS

Assay (Throughout this *Assay*, protect from direct light all solutions containing Natamycin.)
  Mobile Phase Dissolve 3.0 g of ammonium acetate and 1.0 g of ammonium chloride in 760 mL of water, and mix. Add 5.0 mL of tetrahydrofuran and 240 mL of acetonitrile, mix, and filter through a 0.5-µm or finer porosity filter. Make adjustments if necessary to meet the system suitability requirements.
  Standard Preparation Transfer about 20 mg of USP Natamycin Reference Standard, accurately weighed, to a 100-mL volumetric flask. Add 5.0 mL of tetrahydrofuran, and sonicate for 10 min. Add 60 mL of methanol, and swirl to dissolve. Add 25 mL of water, and mix. Allow to cool to room temperature. Dilute with water to volume, mix, and filter through a membrane filter of 5-µm or finer porosity.
  Resolution Solution Dissolve 20 mg of Natamycin in a mixture of 99 mL of methanol and 1 mL of 0.1 *N* hydrochloric acid, and allow to stand for 2 h. (NOTE: Use this solution within 1 h.)
  Assay Preparation Transfer about 20 mg of Natamycin, accurately weighed, to a 100-mL volumetric flask. Proceed as directed under *Standard Preparation*, beginning with "add 5.0 mL of tetrahydrofuran. . . ."
  Chromatographic System (see *Chromatography*, page 471) Use a high-pressure liquid chromatograph equipped with an ultraviolet detector measuring at 303 nm and a 4.6-mm × 25-cm column packed with octadecylsilanized silica (Supelcosil LC 18 or equivalent). The flow rate is about 3 mL/min. Chromatograph the *Standard Preparation*, and record the peak responses: the column efficiency is not less than 3000 theoretical plates, the tailing factor is between 0.8 and 1.3, and the relative standard deviation for three replicate injections is not more than 1.0%. Chromatograph the *Resolution Solution*. The resolution between Natamycin and its methyl ester is not less than 2.5. The relative retention times are about 0.7 for Natamycin and 1.0 for its methyl ester.
  Procedure Separately inject about 20 µL each of the *Standard Preparation* and the *Assay Preparation* into the chromatograph, and record the peak areas of the major peaks. Calculate the percentage of Natamycin in the portion taken by the formula:

$$0.1(W_s P_s/W_u)(r_u/r_s),$$

in which W_s is the weight, in mg, of USP Natamycin Reference Standard taken to prepare the *Standard Preparation*; P_s is the stated content, in µg/mg, of USP Natamycin Reference Standard; W_u is the weight, in mg, of Natamycin taken to prepare the *Assay Preparation*; and r_u and r_s are the peak area responses obtained with the *Assay Preparation* and the *Standard Preparation*, respectively.

Arsenic A *Sample Solution* prepared as directed for organic compounds meets the requirements of the *Arsenic Test*, page 464, using 1 mL of the *Standard Arsenic Solution* (1 µg As).
Heavy Metals Prepare and test a 1-g sample as directed in *Method II* under the *Heavy Metals Test*, page 513, using 20 µg of lead ion (Pb) in the control (*Solution A*).
pH Determine by the *Potentiometric Method*, page 531, using an aqueous suspension containing 10 mg/mL.
Specific Rotation, page 530 Determine in a solution containing 100 mg in each 10 mL of glacial acetic acid.
Water Determine by the *Karl Fischer Titrimetric Method*, page 552.

Functional Use in Foods Antimycotic.
Packaging and Storage Store in tight, light-resistant containers in a cool place.

Insert the following new monographs to precede the monograph entitled *Nutmeg Oil*, page 206:

Nitrogen

N_2 Mol wt 28.01

DESCRIPTION

A colorless and odorless gas. It may be condensed to a colorless liquid boiling at −195.8° or to a white solid melting at −209.8°. It is nonflammable and does not support combustion. One L of the gas weighs about 1.25 g at 0° and a pressure of 760 mm of mercury. One volume of the gas dissolves in about 62 volumes of water and in about 8 volumes of alcohol at 20°.

REQUIREMENTS

NOTE: Reduce the container pressure by means of a regulator. Measure the gas with a gas volume meter downstream from the pertinent detector tube to minimize contamination of or change to the specimens.

Identification

Insert a burning wood splinter into a test tube filled with the gas. The flame is extinguished. (NOTE: Use caution.)

Assay   Not less than 99.0% N_2, by volume.
Carbon Dioxide   Not more than 0.03%, by volume.
Carbon Monoxide   Not more than 10 ppm, by volume.
Oxygen   Not more than 1.0%, by volume.
Water   Passes test.

TESTS

Assay   Introduce a specimen of Nitrogen into a gas chromatograph by means of a gas sampling valve. Select the operating conditions of the gas chromatograph so that the standard peak signal resulting from the following procedure corresponds to not less than 70% of the full-scale reading. Preferably, use an apparatus corresponding to the general type in which the column is 3 m in length and 4 mm in inside diameter and is packed with a molecular sieve prepared from a synthetic alkali-metal aluminosilicate capable of absorbing molecules with diameters of up to 0.5 nm, which permits complete separation of oxygen from Nitrogen. Use industrial-grade helium (99.99%) as the carrier gas, with a thermal conductivity detector, and control the column temperature: the peak response produced by the assay specimen exhibits a retention time corresponding to that produced by the oxygen–helium certified standard (a mixture of 1.0% oxygen in industrial-grade helium is available from most suppliers) and is equivalent to not more than 1.0% of oxygen, by volume, when compared with the peak response of the oxygen–helium certified standard, indicating not less than 99.0% N_2, by volume.

Carbon Dioxide   Pass 1050 ± 50 mL of the gas sample through a carbon dioxide detector tube (described on page 173 of THIS SUPPLEMENT) at the rate specified for the tube. The indicator change corresponds to not more than 0.03%, by volume.

Carbon Monoxide   Pass 1050 ± 50 mL of the gas sample through a carbon monoxide detector tube (described on page 173 of THIS SUPPLEMENT) at the rate specified for the tube. The indicator change corresponds to not more than 10 ppm, by volume.

Oxygen   Determine as directed in the *Assay*.

Water   Pass 24,000 mL of the gas sample through a suitable water-absorption tube no less than 100 mm in length, which previously has been flushed with about 500 mL of the sample and weighed. Regulate the flow so that about 60 min will be required for passage of the gas. The gain in weight of the absorption tube does not exceed 1.0 mg.

Functional Use in Foods   *Gas*: air and oxygen displacer; propellant and aerating agent; *liquid*: direct-contact freezing agent.

Packaging and Storage   Preserve in cylinders.

Nitrogen Enriched Air

N_2 Mol wt 28.01

DESCRIPTION

Nitrogen Enriched Air is produced from air *in situ* by physical separation methods. It contains not less than 90% and not more than 99% nitrogen, by volume. The remaining components are noble gases and, primarily, oxygen.

REQUIREMENTS

NOTE: Reduce the container pressure by means of a regulator. Measure the gas with a gas volume meter downstream from the detector tube to minimize contamination of or change to the specimens.

Identification

Insert a burning wood splinter into a test tube filled with the gas. The flame is extinguished. (NOTE: Use caution.)

Assay   Not less than 90.0% and not more than 99.0% N_2, by volume.
Carbon Dioxide   Not more than 0.03%, by volume.
Carbon Monoxide   Not more than 10 ppm, by volume.
Nitric Oxide and Nitrogen Dioxide   Not more than 2.5 ppm, by volume.

Oxygen Not less than 1.0% and more than 10.0% O_2, by volume.
Sulfur Dioxide Not more than 5 ppm, by volume.
Water Passes test.

ADDITIONAL REQUIREMENTS

Labeling Where the gas is piped from cylinders or directly from the collecting tank to the point of use, label each outlet "Nitrogen Enriched Air."

TESTS

Assay Introduce a specimen of Nitrogen Enriched Air into a gas chromatograph by means of a gas sampling valve. Select the operating conditions of the gas chromatograph so that the peak signal of an oxygen helium certified standard (a mixture of 5.0% oxygen, by volume, in industrial-grade helium is available from most suppliers) resulting from the following procedure corresponds to approximately 45% of the full-scale reading. Preferably, use an apparatus corresponding to the general type in which the column is 3 m in length and 4 mm in inside diameter and is packed with a molecular sieve prepared from a synthetic alkali-metal aluminosilicate capable of absorbing molecules with diameters of up to 0.5 nm, which permits complete separation of oxygen from nitrogen. Use industrial-grade helium (99.99%) as the carrier gas, with a thermal conductivity detector, and control the column temperature: the peak response produced by the assay specimen exhibits a retention time corresponding to that produced by the 5.0% oxygen–helium standard and, when compared with the peak response of the standard, is equivalent to not less than 1.0% and not more than 10.0% of oxygen, indicating not less than 90.0% and not more than 99.0% N_2, by volume.

Carbon Dioxide Pass 1050 ± 50 mL of the gas sample through a carbon dioxide detector tube (described on page 173 of THIS SUPPLEMENT) at the rate specified for the tube. The indicator change corresponds to not more than 0.03%, by volume.

Carbon Monoxide Pass 1050 ± 50 mL of the gas sample through a carbon monoxide detector tube (described on page 173 of THIS SUPPLEMENT) at the rate specified for the tube. The indicator change corresponds to not more than 10 ppm, by volume.

Nitric Oxide and Nitrogen Dioxide Pass 550 ± 50 mL through a nitric oxide–nitrogen dioxide detector tube (described on page 174 of THIS SUPPLEMENT) at the rate specified for the tube. The indicator change corresponds to not more than 2.5 ppm, by volume.

Oxygen Determine as directed in the *Assay*.

Sulfur Dioxide Pass 1050 ± 50 mL through a sulfur dioxide detector tube (described on page 174 of THIS SUPPLEMENT) at the rate specified for the tube. The indicator change corresponds to not more than 5 ppm, by volume.

Water Pass 24,000 mL of the gas sample through a suitable water-absorption tube no less than 100 mm in length, which previously has been flushed with about 500 mL of the sample and weighed. Regulate the flow so that about 60 min will be required for passage of the gas. The gain in weight of the absorption tube does not exceed 1.0 mg.

Functional Use in Foods Air and oxygen displacer.
Packaging and Storage Preserve in metal cylinders or in a low-pressure collecting tank.

Nitrous Oxide

Nitrogen Oxide (N_2O)

N_2O Mol wt 44.01

DESCRIPTION

A colorless gas without appreciable odor. One L at 0° and at a pressure of 760 mm of mercury weighs about 1.97 g. One volume dissolves in about 1.4 volumes of water at 20° and at a pressure of 760 mm of mercury. It is freely soluble in alcohol and soluble in ether and in oils.

REQUIREMENTS

NOTE: The following tests are designed to reflect the quality of Nitrous Oxide in both its vapor and liquid phases, which are present in previously unopened cylinders. Reduce the container pressure by means of a regulator. Withdraw the samples for the tests with the least possible release of Nitrous Oxide consistent with proper purging of the sample apparatus. Measure the gases with a gas volume meter downstream from the detector tubes to minimize contamination of or change to the specimens. Perform tests in the sequence in which they are listed under *Requirements*.

Identification

A. With the container temperatures the same and maintained between 15° and 25°, concomitantly read the pressure of the Nitrous Oxide container and of a container of 99.9% Nitrous Oxide certified standard (available from most suppliers). (NOTE: Do not use the Nitrous Oxide certified standard if it has been depleted to less than half of its full capacity.) The pressure of the Nitrous Oxide container is within 50 psi of that of the Nitrous Oxide certified standard.

B. Pass 100 ± 5 mL released from the vapor phase of the contents of the Nitrous Oxide container through a carbon dioxide detector tube (described on page 173 of THIS

SUPPLEMENT) at the rate specified for the tube: no color change is observed (distinction from carbon dioxide).

Carbon Monoxide Not more than 10 ppm, by volume.
Nitric Oxide Not more than 1 ppm, by volume.
Nitrogen Dioxide Not more than 1 ppm, by volume.
Halogens (as Cl) Not more than 1 ppm, by volume.
Carbon Dioxide Not more than 0.03%, by volume.
Ammonia Not more than 0.0025%, by volume.
Water Not more than 150 mg/m^3.
Odor Passes test.
Air Not more than 1.0%, by volume.
Assay Not less than 99.0% N_2O, by volume.

TESTS

Assay Introduce a specimen of Nitrous Oxide taken from the liquid phase, as directed in the test for *Nitrogen Dioxide*, into a gas chromatograph by means of a gas sampling valve. Select the operating conditions of the gas chromatograph such that the peak response resulting from the following procedure corresponds to not less than 70% of the full-scale reading. Preferably, use an apparatus corresponding to the general type in which the column is 6 m in length and 4 mm in inside diameter and is packed with porous polymer beads, which permit complete separation of nitrogen and oxygen from Nitrous Oxide, although the nitrogen and oxygen may not be separated from each other. Use industrial-grade helium (99.99%) as the carrier gas, with a thermal-conductivity detector, and control the column temperature: the peak response produced by the assay specimen exhibits a retention time corresponding to that produced by an air–helium certified standard (described on page 119 of THIS SUPPLEMENT) and is equivalent to not more that 1.0% of air, by volume, when compared with the peak response of the air–helium certified standard, indicating not less than 99.0% N_2O, by volume.

Air Determine as directed in the *Assay*.

Ammonia Pass 1050 ± 50 mL, released from the vapor phase of the contents of the container, through an ammonia detector tube (described on page 173 of THIS SUPPLEMENT) at the rate specified for the tube: the indicator change corresponds to not more than 0.0025%, by volume.

Carbon Dioxide Pass 1050 ± 50 mL, released from the vapor phase of the contents of the container, through a carbon dioxide detector tube at the rate specified for the tube: the indicator change corresponds to not more than 0.03%, by volume.

Carbon Monoxide Pass 1050 ± 50 mL, released from the vapor phase of the contents of the container, through a carbon monoxide detector tube (described on page 173 of THIS SUPPLEMENT) at the rate specified for the tube: the indicator change corresponds to not more than 10 ppm, by volume.

Halogens Pass 1050 ± 50 mL, released from the vapor phase of the contents of the container, through a chlorine detector tube (described on page 173 of THIS SUPPLEMENT) at the rate specified for the tube: the indicator change corresponds to not more than 1 ppm, by volume.

Nitric Oxide Pass 550 ± 50 mL, released from the vapor phase of the contents of the container, through a nitric oxide–nitrogen dioxide detector tube (described on page 174 of THIS SUPPLEMENT) at the rate specified for the tube: the indicator change corresponds to not more than 1 ppm, by volume.

Nitrogen Dioxide Arrange a container so that when its valve is opened, a portion of the liquid phase of the contents is released through a piece of tubing of sufficient length to allow all of the liquid to vaporize during passage through it and to prevent frost from reaching the inlet of the detector tube. Release into the tubing a flow of liquid sufficient to provide 550 mL of the vaporized sample plus any excess necessary to ensure adequate flushing of air from the system. Pass 550 ± 50 mL of this gas through a nitric oxide–nitrogen dioxide detector tube at the rate specified for the tube: the indicator change corresponds to not more than 1 ppm, by volume.

Odor Carefully open the container valve to produce a moderate flow of gas. Do not direct the gas stream toward the face, but deflect a portion of the stream toward the nose: no appreciable odor is discernible.

Water Flush the regulator with 5 L or more of the gas specimen. Pass 50 ± 5 L, released from the vapor phase, through a water vapor detector tube (described on page 174 of THIS SUPPLEMENT) connected to the regulator with a minimum length of metal or polyethylene tubing. Measure the gas passing through the detector tube with a gas flowmeter set at a flow rate of 2 L/min. The corrected indicator change corresponds to not more than 150 mg/m^3.

Functional Use in Foods Propellant; aerating agent; gas.
Packaging and Storage Preserve in cylinders.

Insert the following new monograph to precede the monograph entitled *Oxystearin*, page 211:

Ox Bile Extract

Sodium Choleate, Purified Oxgall

$C_{24}H_{39}NaO_5$ Formula wt 430.57 (as sodium cholate)

DESCRIPTION

A yellowish green powder with a partly sweet, partly bitter, disagreeable taste. It is soluble in water and in alcohol. It contains ox bile acids, chiefly glycocholic and taurocholic, as sodium salts, equivalent to not less than 45.0% cholic acid ($C_{24}H_{40}O_5$). It is the purified portion of the bile of an ox, obtained by evaporating the alcohol extract of concentrated bile.

REQUIREMENTS

Identification

A. It gives a positive test for *Sodium*, page 517.
B. A 1 in 10 solution in alcohol, when mixed with 0.5 mL of iodine TS, yields a blue color.

Assay   Contains the equivalent of not less than 45.0% of cholic acid ($C_{24}H_{40}O_5$).
Ash (Total)   Not more than 10.0%.
Loss on Drying   Not more than 6.0%.
Microbial Limits:
  Aerobic Plate Count   Not more than 20,000 per g.
  E. coli   Not more than 3 per g.
  Salmonella sp.   Negative by test.
  Yeasts and Mold   Not more than 10 per g.
pH   Between 6.3 and 7.0.

TESTS

Assay
  Standard Solution   Using an accurately weighed quantity of USP Cholic Acid Reference Standard, prepare a solution in 60% acetic acid with a known concentration of about 0.5 mg/mL. When stored in a refrigerator, this solution may be used for several months.
  Assay Preparation   Dissolve about 50 mg of the sample, accurately weighed, in 100 mL of 60% acetic acid, and mix. Filter the solution, discarding the first 10 mL of the filtrate.
  Procedure   Transfer 1.0 mL each of the *Standard Solution* and the *Assay Preparation* into separate containers. To each container, add 1.0 mL of a freshly prepared 1 in 100 solution of furfural. Cool the containers in an ice bath for 5 min. Add 13 mL of a dilute solution of sulfuric acid, prepared by cautiously mixing 50 mL of sulfuric acid with 65 mL of water. Thoroughly mix the contents in each container, and place them in a water bath maintained at 70° for 10 min. Then immediately place the containers in an ice bath for 2 min. Determine the absorbance of each solution in a 1-cm cell at the wavelength of maximum absorbance at about 650 nm. Calculate the quantity, in mg, of cholic acid taken in the portion of Ox Bile Extract by the formula:

$$100C(A_U/A_S),$$

in which C is the concentration, in mg/mL, of USP Cholic Acid Reference Standard in the *Standard Solution*, and A_U and A_S are the maximum absorbances at identical wavelengths of the *Assay Preparation* and the *Standard Solution*, respectively.

Ash (Total)   Determine by the general method, page 466.
Loss on Drying, page 518   Dry at 105° for 16 h.
Microbial Limits:
  Aerobic Plate Count   Proceed as directed in Chapter 4 of FDA's *Bacteriological Analytical Manual*, 6th ed., Second Printing, 1989.
  E. coli   Proceed as directed in Chapter 6 of FDA's *Bacteriological Analytical Manual*, 6th ed., Second Printing, 1989.
  Salmonella sp.   Proceed as directed in Chapter 7 of FDA's *Bacteriological Analytical Manual*, 6th ed., Second Printing, 1989.
  Yeasts and Molds   Proceed as directed in Chapter 19 of FDA's *Bacteriological Analytical Manual*, 6th ed., Second Printing, 1989.
pH   Determine by the *Potentiometric Method*, page 531, of a 1 in 20 solution.

Functional Use in Foods   Surfactant.
Packaging and Storage   Store in tight containers.

Insert the following new monograph to precede the monograph entitled *Palmarosa Oil*, page 212:

Ozone

Triatomic Oxygen

O_3 Mol wt 47.9982

DESCRIPTION

Ozone is an unstable, colorless gas with a pungent, characteristic odor. It is produced *in situ* from oxygen either by ultraviolet irradiation of air or by passing a high-voltage discharge through air. It is a potent oxidizing agent that decomposes at ambient temperature to molecular oxygen.

REQUIREMENTS

Identification

  Reagent Solution   Disperse 124.5 mg of alizarin violet 3R in 500 mL of water in a 1-L volumetric flask. Mechanically stir overnight. Add 20 mg of sodium hexametaphosphate, 48.5 g of ammonium chloride, and 6.2 mL of ammonium hydroxide (equivalent to 1.6 g of ammonia). Dilute to volume with water, and stir overnight. A 10-fold dilution of this solution has an absorbance of 0.155 AU cm^{-1} at 548 nm; the pH of dilutions with sample waters is between 8.1 and 8.5.

Procedure Introduce 20 mL of the *Reagent Solution* into each of two 200-mL volumetric flasks. Fill one flask with Ozone-free water to serve as the blank. Fill the other with the sample by directly introducing the sample, with the aid of a long-stemmed funnel or pipet, below the surface of the *Reagent Solution* to prevent Ozone loss by degassing. Immediately measure the absorbance of both solutions at 548 nm, using 1- to 5-cm cells: the presence of Ozone is indicated if the sample solution has a lower absorbance than the blank.

Assay 0.01 to 0.5 mg of O_3/L.

TESTS

Assay

Indigo Stock Solution In a 1-L volumetric flask, dissolve 0.770 g of potassium indigotrisulfonate in 500 mL of water and 1 mL of phosphoric acid, dilute to volume with water, and mix. A 1:100 dilution of this reagent has an absorbance of 0.20 ± 0.010 cm^{-1} at 600 nm.

Indigo Reagent I Just before use, transfer 20 mL of *Indigo Stock Solution*, 10 g of monobasic sodium phosphate, and 7 mL of phosphoric acid into a 1-L volumetric flask, dilute to volume with water, and mix.

Indigo Reagent II Proceed as directed for *Indigo Reagent I*, using 100 mL of *Indigo Stock Solution* instead of 20 mL.

Malonic Acid Reagent Dissolve 5 g of malonic acid in water and dilute to 100 mL.

Procedure (for a concentration range of 0.01 to 0.1 mg Ozone/L) Add 10.0 mL of *Indigo Reagent I* to each of two 100-mL flasks. Fill one flask with Ozone-free water to serve as the blank. Fill the other with the sample by directly introducing the sample, with the aid of a long-stemmed funnel or pipet, below the surface of the dye solution to prevent Ozone loss by degassing. Without delay, mix and measure the absorbance of each solution at 600 nm, preferably in 10-cm cells. (For a concentration range of 0.05 to 0.5 mg of Ozone/L, use *Indigo Reagent II* and proceed as above.)

Control of Interferences In the presence of chlorine, add 1 mL of *Malonic Acid Reagent* to both flasks before adding the samples. Proceed as above, but measure absorbance immediately.

Calculate the concentration of ozone, in mg/L, by the formula:

$$100D/(f \times b \times V),$$

in which D is the difference in absorbance between the sample solution and blank solution; b is the path length, in cm; V is the volume of sample, in mL (normally 90 mL); and f is 0.42.

Functional Use in Foods Antimicrobial and disinfectant for water to be used for direct consumption, such as for ice, or for indirect consumption, such as for water used in the treatment or display of fish, produce, and other perishable foods. It is also used in the treatment of wastewater.

Palmarosa Oil, page 212

Change the *Requirements* entitled *Assay for Alcohols* and *Refractive Index* to read:

Assay for Alcohols Not less than 88.0% total alcohols.
Refractive Index Between 1.470 and 1.476 at 20°.

Pectin, page 215

Change the monograph title to read:

Pectins

Change the *Description* to read:

Pectins are a purified heterogeneous complex carbohydrate polymer mixture of varied molecular weights generally obtained by dilute-acid extraction of citrus albedo or apple pomace. They usually occur as practically odorless, yellowish white, coarse to fine powders with a mucilaginous taste. They dissolve in water, forming an opalescent colloidal solution. They are practically insoluble in alcohol.

The major part of the pectin chain is composed of $\alpha(1 \rightarrow 4)$-linked D-galacturonic acid units. Some of the carboxyl groups are esterified with methyl alcohol, while the remaining carboxylic units exist in the free acid form or as ammonium, potassium, or sodium salts. A portion of the methyl esters may have been converted to primary amides. Commercial Pectin products are normally diluted with sugar for standardization. In addition to sugars, suitable food-grade buffer salts may be added for pH control and to achieve desirable setting characteristics.

NOTE: The following *Requirements* and *Tests* apply to the Pectins as supplied, whether standardized or not (except for specifications covering amide substitution) and the weight percent of total galacturonides in the Pectin component, in which cases the test procedures provide for removing the sugars and soluble salts before analysis of the Pectin component.

Change *Identification Tests*, *A*, *B*, and *C*, to read:

A. To a 1 in 100 aqueous solution of the sample add an equal volume of alcohol. A translucent, gelatinous precipitate is formed (most gums will not form such a precipitate).
B. To 10 mL of a 1 in 100 aqueous solution of the sample add 1 mL of thorium nitrate solution (1 in 10), stir, and allow to stand for 2 min. A stable precipitate or gel forms (most gums will not form such a precipitate).
C. To 5 mL of a 1 in 100 aqueous solution of the sample add 1 mL of sodium hydroxide TS, and allow to stand at room temperature for 15 min. A gel or semigel forms (tragacanth and other gums will not form such a precipitate).

Delete the *Requirements* for *Ash (Total)*, *Degree of Esterification of High-Ester Pectin Component*, and *Degree of Esterification of Low-Ester Pectin Component*.

Change the name of the *Requirement* entitled *Degree of Amide Substitution of Low-Ester Pectin Component* to read:

Degree of Amide Substitution

Change the *Requirements* entitled *Degree of Amide Substitution*, *Heavy Metals*, *Lead*, *Loss on Drying*, and *Total Anhydrogalacturonides in Pectin Component* to read:

Degree of Amide Substitution Not more than 25% total carboxylic groups.
Heavy Metals (as Pb) Not more than 20 mg/kg.
Lead Not more than 5 mg/kg.
Loss on Drying Not more than 12.0%.
Total Anhydrogalacturonides in Pectin Component Not less than 65.0%.

Insert the following under *Requirements*:

Methanol, Ethanol, and Isopropanol Not more than 1% total.
Sulfur Dioxide Not more than 0.005%.

Delete the *Tests* entitled *Ash (Total)* and *Degree of Esterification*.

Change the *Tests* entitled *Degree of Amide Substitution* and *Total Anhydrogalacturonides in the Pectin Component*, *Heavy Metals*, *Lead*, and *Sodium Methyl Sulfate* to read:

Degree of Amide Substitution and **Total Anhydrogalacturonides in the Pectin Component** Transfer 5.0 g of the sample to a beaker, and stir for 10 min with a mixture of 5 mL of hydrochloric acid and 100 mL of 60% isopropyl alcohol. Filter through a dry, coarse sintered-glass filter tube (30- to 60-mL capacity), and wash with six 15-mL portions of the acid–alcohol mixture and then with 60% isopropyl alcohol until the filtrate is free from chloride.

Finally, wash with 20 mL of anhydrous isopropyl alcohol, dry at 105° for 2.5 h, and cool.

Transfer 500 mg, accurately weighed, of the washed and dried sample into a 250-mL Erlenmeyer flask, and moisten with 2 mL of alcohol. Add 100 mL of carbon dioxide-free water, stopper, and swirl occasionally until the sample is completely hydrated. Add 5 drops of phenolphthalein TS, and titrate with 0.1 N sodium hydroxide, recording the volume, in mL, required as V_1 (initial titer). Add 20.0 mL of 0.5 N sodium hydroxide, stopper, shake vigorously, and allow to stand for 15 min. Add 20.0 mL of 0.5 N hydrochloric acid, shake until the pink color disappears, and then add 3 drops of phenolphthalein TS, and titrate with 0.1 N sodium hydroxide to a faint pink color that persists after vigorous shaking. Record the volume, in mL, of 0.1 N sodium hydroxide required as V_2 (saponification titer).

Quantitatively transfer the contents of the flask into a 500-mL distillation flask fitted with a Kjeldahl trap and a water-cooled condenser, the delivery tube of which extends well beneath the surface of a mixture of 150 mL of carbon dioxide-free water and 20.0 mL of 0.1 N hydrochloric acid contained in a receiving flask. To the distillation flask add 20 mL of sodium hydroxide solution (1 in 10), seal the connections, and begin heating carefully to avoid excessive foaming. Continue heating until 80 to 120 mL of distillate has been collected. Disconnect the receiver. Add a few drops of methyl red TS to the receiving flask, and titrate the excess acid with 0.1 N sodium hydroxide, recording the volume, in mL, required as S. Perform a blank determination on 20.0 mL of 0.1 N hydrochloric acid, and record the volume, in mL, required as B. Record the amide titer $(B - S)$ as V_3, and the total titer $(V_1 + V_2 + V_3)$ as V_t. (NOTE: If the Pectin is known to be of the nonamidated type, only V_1 and V_2 need be determined and V_3 may be regarded as zero.)

Calculate the degree of amide substitution by the formula $100(V_3/V_t)$. Calculate the weight percent of total anhydrogalacturonides by the formula $3.52V_1 + 3.80V_2 + 3.5V_3$.

Heavy Metals Prepare and test a 1-g sample as directed in *Method II* under the *Heavy Metals Test*, page 513, using 20 µg of lead ion (Pb) in the control (*Solution A*).

Lead

Lead Nitrate Stock Solution Dissolve 159.8 mg of lead nitrate in 100 mL of water containing 1 mL of nitric acid (alternatively, use NIST Standard Reference Material, containing 10 mg of lead per kg, or equivalent). Dilute to 1000.0 mL with water, and mix. Prepare and store this solution in glass containers that are free from lead salts. Each mL of this solution contains the equivalent of 100 µg of lead ion (Pb).

Standard Lead Solutions Immediately before use, prepare a series of lead standard solutions serially diluted from the *Lead Nitrate Stock Solution* in water. Into separate 100-

mL volumetric flasks, pipet 0.0, 0.5, 1.0, 2.0, 5.0, and 10.0 mL, respectively, of *Lead Nitrate Stock Solution*, dilute to volume with water, and mix. The *Standard Lead Solutions* contain, respectively, 0.0, 0.5, 1.0, 2.0, 5.0, and 10.0 µg of lead per mL.

Sample Preparation (NOTE: Perform this procedure in a fume hood and wear safety glasses.) Transfer about 2 g of the sample, accurately weighed, into a 50-mL Erlenmeyer flask. Add 5 mL of sulfuric acid with a few glass beads, and heat on a hot plate at 110°. An additional volume, not to exceed 5 mL of sulfuric acid, may be needed to completely digest the sample. While heating, add 30% hydrogen peroxide dropwise, allowing the reaction to subside and reheating between the addition of drops. The first few drops of hydrogen peroxide should be added cautiously, and heating should be discontinued if excessive foaming occurs. Swirl the solution in the flask to prevent the unreacted sample from caking on the walls of the flask. Add small quantities of hydrogen peroxide when the sample begins to darken, and continue the digestion process until all of the organic matter is destroyed. Slowly raise the temperature of the hot plate to 250°, and allow fumes to evolve from the sample until it becomes colorless or retains a light straw color. Cool the solution. Cautiously add 10 mL of water, again evaporate almost to dryness, and cool. Add 5 mL of 1 *N* hydrochloric acid to the sample and quantitatively transfer it into a 10-mL volumetric flask, dilute to volume with 1 *N* hydrochloric acid, and mix. Prepare a reagent blank solution in the same manner.

Procedure Concomitantly determine the absorbances of the *Standard Lead Solutions* and the *Sample Preparation* at the lead emission line of 283.3 nm, with a suitable atomic absorption spectrophotometer equipped with a lead hollow-cathode lamp and an oxidizing air–acetylene flame, using the reagent blank solution to zero the instrument. Plot the absorbances of the *Standard Lead Solutions* versus the concentration, in µg/mL, of lead, and draw the straight line that best fits the plotted points. From the graph so obtained, determine the concentration C, in µg/mL, of the *Sample Preparation*. Calculate the quantity, in mg/kg, of lead in the sample by the formula:

$$10C/W,$$

in which W is the weight, in g, of the sample taken.

Sodium Methyl Sulfate

Mobile Phase Prepare a 0.04 *M* potassium hydrogen phthalate solution by transferring 16.4 g of potassium hydrogen phthalate into a 2-L volumetric flask, dilute to volume with water, and mix. Filter the solution through a 0.45-µm pore-size filter (Millipore® or equivalent).

Standard Preparation Transfer 10.0 mg of anhydrous sodium methyl sulfate into a 100-mL volumetric flask, dilute to volume with *Mobile Phase*, and mix.

Assay Preparation Suspend about 1 g of the sample, accurately weighed, in 10.0 mL of 50% (v/v) ethanol solution. Stir for 30 min using a Teflon-coated stirring bar. Allow the suspension to precipitate, and filter. Evaporate a 1.0-mL aliquot to dryness using reduced pressure (10 mm Hg), and heat at 60°. Redissolve the residue in 1.0 mL of the *Mobile Phase*.

Chromatographic System Use a high-performance liquid chromatograph equipped with a refractive index detector and a 250-mm × 4.6-mm column packed with Nucleosil 10SB (or equivalent) and maintained at 40°. Set the flow rate at 1 mL/min.

System Suitability Three replicate injections of the *Standard Preparation* show a relative standard deviation of not more than 4.0% for the response factor of the sodium methyl sulfate peak obtained using the formula (A_s/C_s), in which A_s is the peak area response of the *Standard Preparation*, and C_s is the concentration, in mg/mL, of sodium methyl sulfate in the *Standard Preparation*.

Procedure Inject 20 µL of the *Standard Preparation* followed by the *Assay Preparation*. Determine the peak area in the chromatograms for the *Standard Preparation* and *Assay Preparation*. Calculate the quantity in percent of sodium methyl sulfate in the sample by the formula:

$$(C_S A_U)/(A_S W),$$

in which C_S is the concentration, in mg/mL, of sodium methyl sulfate in the *Standard Preparation*; A_U and A_S are the peak area responses obtained from the *Assay Preparation* and *Standard Preparation*, respectively; and W is the weight, in g, of the sample taken.

Insert the following under *Tests*:

Methanol, Ethanol, and Isopropanol The alcohols are converted to their nitrite esters, and their levels are determined by headspace gas chromatography.

Internal Standard Solution Dissolve 50 mg of *n*-propanol in 1 L of water.

Sample Solution Dissolve 100 mg of the sample in 10 mL of water, and as necessary, use sodium chloride as a dispersing agent.

Standard Alcohol Solution Pipet 50 mg each of methanol (corresponding to 39.55 mL), ethanol (corresponding to 39.47 mL), and isopropanol (corresponding to 39.28 mL) into a 1000-mL volumetric flask, dilute to volume with water, and mix.

Sodium Nitrite Solution Dissolve 250 g of sodium nitrite in 1000 mL of water.

Chromatographic System Use a suitable gas chromatograph equipped with a flame-ionization detector. Use a 90-cm × 4-mm i.d. glass column with the first 15 cm packed with Chromopack (or equivalent) and the remainder packed with Poropak R 120–150 mesh (or equivalent). The operating conditions of the gas chromatograph are as follows: the injection port temperature is 250°, and the column temperature is 150° isothermal. Use nitrogen as the carrier gas with a flow rate of 80 mL/min.

Procedure   Weigh 200 mg of urea, and place it in a 25-mL amber-glass vial (Reacti-Flasks or equivalent). Purge with nitrogen for 5 min, add 1 mL of saturated oxalic acid solution, close with a rubber stopper, and swirl. Add 1 mL of *Sample Solution* and 1 mL of *Internal Standard Solution*, and simultaneously start a stopwatch ($t = 0$). Swirl the vial, and recap it with an open screw cap fitted with a silicone rubber septum. Swirl the vial until $t = 30$ s. At $t = 45$ s, inject through the septum 0.5 mL of *Sodium Nitrite Solution*. Swirl until $t = 70$ s, and at $t = 150$ s, withdraw through the septum 1.0 mL of the headspace using a pressure lock syringe (Precision Sampling Corporation or equivalent). Inject the 1.0 mL into the injection port of the gas chromatograph. Repeat this procedure but use 1 mL of the *Standard Alcohol Solution* instead of the *Sample Solution*.

Calculation   Quantify the total methanol, ethanol, and isopropanol present in the sample by the following formula:

$$T = V_{MS}(R_{MU}/R_{MS})0.791 + V_{ES}(R_{EU}/R_{ES})0.7893 + V_{IS}(R_{IU}/R_{IS})0.7855,$$

in which T is the total amount, in mg, of methanol, ethanol, and isopropanol in the sample; the subscripts M, E, and I refer to methanol, ethanol and isopropanol, respectively; V_S is the volume, in mL, of the corresponding alcohol in the *Standard Alcohol Solution*; R_S is the ratio of the peak area of the corresponding alcohol in the *Standard Alcohol Solution* to that of *n*-propanol in *Internal Standard Solution*; R_U is the ratio of the peak area of the corresponding alcohol in the *Sample Solution* to that of *n*-propanol in the *Internal Standard Solution*; and 0.791, 0.7893, and 0.7855 are the densities, in g/mL, for methanol, ethanol, and isopropanol, respectively. Calculate the percent methanol, ethanol, and isopropanol present in the sample by the following formula:

$$(T100)/W,$$

in which W is the sample weight, in mg.

Sulfur Dioxide   Determine as directed in the general method, page 170 of THIS SUPPLEMENT, using the following method under *Sample Introduction and Distillation*. Transfer about 20 g of the sample, accurately weighed, into flask C, and add 20 mL of ethanol to moisten the sample. Add 400 mL of water, swirling vigorously to disperse the sample. Reassemble the apparatus, making sure that the tapered joints are clean and greased with stopcock grease, and proceed as directed under *Sample Introduction and Distillation* beginning with "the nitrogen flow through the 3% hydrogen peroxide solution. . . ."

Perlite, page 219

Insert the following as the penultimate sentence in the *Description*:

Acceptable food-grade free-flowing agents such as sodium carbonate and sodium silicate may be added.

Change the *Requirement* entitled *pH* to read:

pH   Between 5 and 11 (filtrate from a 10% suspension).

DL-Phenylalanine, page 223

Change the *Requirements* entitled *Assay* and *Arsenic* to read:

Assay   Not less than 98.5% and not more than 101.5% $C_9H_{11}NO_2$, calculated on the dried basis.
Arsenic (as As)   Not more than 1.5 mg/kg.

Delete the *Requirements* entitled *Ammonium Salts*, *Chloride*, *Iron*, *Nitrogen*, and *Sulfate*.

Delete the *Tests* entitled *Ammonium Salts*, *Chloride*, *Iron*, *Nitrogen*, and *Sulfate*.

Change the *Test* entitled *Arsenic* to read:

Arsenic   A *Sample Solution* prepared as directed for organic compounds meets the requirements of the *Arsenic Test*, page 464, using 1.5 mL of the *Standard Arsenic Solution* in the control (1.5 µg As).

L-Phenylalanine, page 224

Change the *Requirements* entitled *Assay*, *Arsenic*, and *Specific Rotation* to read:

Assay   Not less than 98.5% and not more than 101.5% $C_9H_{11}NO_2$, calculated on the dried basis.
Arsenic (as As)   Not more than 1.5 mg/kg.
Specific Rotation $[\alpha]_D^{20°}$: Between $-33.2°$ and $-35.2°$, calculated on the dried basis.

Change the *Test* entitled *Arsenic* to read:

Arsenic   A *Sample Solution* prepared as directed for organic compounds meets the requirements of the *Arsenic Test*, page 464, using 1.5 mL of the *Standard Arsenic Solution* in the control (1.5 µg As).

Polydextrose, Second Supplement, page 57

Change the *Requirement* entitled *Heavy Metals* to read:

Heavy Metals (as Pb)   Not more than 5 mg/kg.

Insert the following under *Requirements*:

Lead   Not more than 0.5 mg/kg.

Replace the *Additional Requirement* entitled *Polydextrose (untreated), pH of a 10% Solution* with the following:

Not less than 2.5.

Change the *Test* entitled *Heavy Metals* to read:

Heavy Metals   Prepare and test a 4.0-g sample as directed in *Method II* under the *Heavy Metals Test*, page 513, using 20 µg of lead ion (Pb) in the control (*Solution A*).

Insert the following under *Tests*:

Lead   Determine as directed under *Method I* in the *Atomic Absorption Spectrophotometric Graphite Furnace Method* under the *Lead Limit Test*, page 168 of THIS SUPPLEMENT, except for the following changes.
   Apparatus   Use a suitable spectrophotometer (Perkin-Elmer Model 6000 or equivalent), a graphite furnace containing a L'vov platform (Perkin-Elmer Model HGA-500 or equivalent), and an autosampler (Perkin-Elmer Model AS-40 or equivalent).   Use a lead hollow cathode lamp (lamp current of 10 mA), a slit width of 0.7 mm (set low), the wavelength set at 283.3 nm, and a deuterium arc lamp for background correction.
   Standard Lead Solution   Dilute to volume with water instead of the *Hydrogen Peroxide–Nitric Acid Solution*, and mix.
   Standard Solutions   Dilute to volume with water instead of the *Hydrogen Peroxide–Nitric Acid Solution*, and mix.
   Matrix Modifier   Transfer 100.0 mg of ammonium phosphate, dibasic $((NH_4)_2HPO_4)$ to a 10-mL volumetric flask, dilute to volume with water, and mix.
   Sample Solution   Transfer about 1 g of the sample, accurately weighed, to a 10-mL volumetric flask, add 5 mL of water, and mix. Dilute to volume with water instead of the *Hydrogen Peroxide–Nitric Acid Solution*, and mix.
   Spiked Sample Solution   Prepare a solution as directed under *Sample Solution*, but add 100 µL of the *Standard Lead Solution*, dilute to volume with water, and mix.   This solution contains 0.1 µg/mL of added lead.
   Procedure   With the use of an autosampler, atomize 10-µL aliquots of the four *Standard Solutions*, using the following sequence of conditions: step (1) dry at 130° with a 20-s ramp period, a 40-s hold time, and a 300-mL/min argon flow rate; step (2) then char at 800° with a 20-s ramp period, a 40-s hold time, and a 300-mL/min argon flow rate; step (3) atomize at 2400° for 6 s with a 50-mL/min argon flow rate, and read; step (4) clean at 2600° with a 1-s ramp period, a 5-s hold time, and a 300-mL/min argon flow rate; and step (5) recharge at 20° with a 2-s ramp period, a 20-s hold time, and a 300-mL/min argon flow rate. Atomize 10 µL of the *Matrix Modifier* in combination with either 10 µL of the *Sample Solution* or 10 µL of the *Spiked Sample Solution* under identical conditions used for the *Standard Solutions*.
   Plot a standard curve using the concentration, in µg/mL, of each *Standard Solution* versus its maximum absorbance value compensated for background correction, and draw the best straight line. From the *Standard Curve*, determine the concentrations C_s and C_a, in µg/mL, of the *Sample Solution* and the *Spiked Sample Solution*, respectively. Calculate the quantity, in mg/kg, of lead in the sample by the formula:

$$10C_s/W,$$

in which W is the weight, in g, of the sample taken. Calculate the recovery by the formula:

$$[(C_s - C_a)/0.1]100,$$

in which 0.1 is the amount of lead, in µg/mL, added to the *Spiked Sample Solution*.

Change the fifth sentence of the *Test* entitled *Monomers, Procedure*, to read:

Calculate the percentage of each monomer (P_M) by the formula:

$$P_M = (R \times W_S)/(R_S \times W),$$

in which W is the weight, in mg, of the sample, adjusted for ash and moisture; W_S is the weight, in mg, of the respective monomer in the *Standard Solution*; R is the ratio of the area of the monomer peak to the area of the octadecane peak in the sample injection; and R_S is the mean ratio of the area of the monomer peak to the area of the octadecane peak in the standard injections.

Insert the following new monograph to precede the monograph entitled *Potassium Bicarbonate*, page 239:

Potassium Benzoate

$C_7H_5KO_2$ Formula wt 160.22

DESCRIPTION

White, odorless or nearly odorless granules, crystalline powder, or flakes. One g dissolves in 2 mL of water, in 75 mL of alcohol, and in 50 mL of 90% alcohol.

REQUIREMENTS

Identification

A 1 in 5 solution responds to the flame test for *Potassium*, page 517, and gives positive tests for *Benzoate*, page 516.

Assay Not less than 99.0% $C_7H_5KO_2$, calculated on the dried basis.
Alkalinity (as KOH) Not more than 0.06%.
Arsenic (as As) Not more than 3 mg/kg.
Heavy Metals (as Pb) Not more than 10 mg/kg.
Water Not more than 1.5%.

TESTS

Assay Transfer about 600 mg, accurately weighed, to a 250-mL beaker, add 100 mL of glacial acetic acid, and stir until the sample is completely dissolved. Add crystal violet TS, and titrate with 0.1 N perchloric acid in glacial acetic acid. Each mL of 0.1 N perchloric acid is equivalent to 16.02 mg of $C_7H_5KO_2$.
Alkalinity Dissolve 2 g in 20 mL of hot water, and add 2 drops of phenolphthalein TS. If a pink color is produced, not more than 0.2 mL of 0.1 N sulfuric acid is required to discharge it.
Arsenic A *Sample Solution* prepared as directed for organic compounds meets the requirements of the *Arsenic Test*, page 464.
Heavy Metals Dissolve 4 g in 40 mL of water, add dropwise, with vigorous stirring, 10 mL of diluted hydrochloric acid TS, and filter. A 25-mL portion of the filtrate meets the requirements of the *Heavy Metals Test*, page 512, using 20 µg of lead ion (Pb) in the control (*Solution A*).
Water Determine by the *Karl Fischer Titrimetric Method*, page 552.

Functional Use in Foods Preservative; antimicrobial agent.
Packaging and Storage Store in well-closed containers.

Potassium Bicarbonate, page 239

Change the *Requirement* entitled *Assay* to read:

Assay Not less than 99.0% and not more than the equivalent of 101.5% $KHCO_3$, calculated on the dried basis.

Potassium Chloride, page 241

Change the *Requirements* entitled *Heavy Metals* and *Loss on Drying* to read:

Heavy Metals (as Pb) Not more than 5 mg/kg.
Loss on Drying Not more than 1.0%.

Change the name of the *Requirement* entitled *Iodide or Bromide* to read:

Iodide and/or Bromide

Change the name of the *Test* entitled *Iodide or Bromide* to read:

Iodide and/or Bromide

Change the *Test* entitled *Heavy Metals* to read:

Heavy Metals Prepare and test a 4-g sample as directed under the *Heavy Metals Limit Test*, page 512, using 20 µg of lead ion (Pb) in the control (*Solution A*).

Insert the following new monograph to precede the monograph entitled *Potassium Metabisulfite*, page 247:

Potassium Lactate Solution

Propanoic Acid, 2-Hydroxy-, Monopotassium Salt

$C_3H_5KO_3$ Formula wt 128.17

DESCRIPTION

Clear, colorless, or practically colorless, viscous liquid, odorless or having a slight, not unpleasant, odor. It is miscible with water. It is available as solutions with concentrations ranging from about 50% to 70%.

REQUIREMENTS

Identification

It gives positive tests for *Potassium*, page 517, and for *Lactate*, page 517.

Assay Not less than 50.0%, by weight, and not less than 95.0% and not more than 105.0%, by weight, of the labeled amount of potassium lactate, $C_3H_5KO_3$.
Chloride Not more than 0.05%.
Citrate, Oxalate, Phosphate, or Tartrate Passes test.

Cyanide Not more than 0.5 mg/kg.
Heavy Metals (as Pb) Not more than 10 mg/kg.
Lead Not more than 5 mg/kg.
Methanol and Methyl Esters Not more than 0.025%.
pH Between 5.0 and 9.0.
Sodium Not more than 0.1%.
Sugars Passes test.
Sulfate Not more than 0.005%.

ADDITIONAL REQUIREMENTS

Labeling Label to indicate its content, by weight, of potassium lactate, $C_3H_5KO_3$.

TESTS

Assay Weigh accurately into a suitable flask a volume of Potassium Lactate Solution equivalent to about 500 mg of potassium lactate, add 60 mL of a 1 in 5 mixture of acetic anhydride in glacial acetic acid, mix, and allow to stand for 20 min. Titrate with 0.1 N perchloric acid in glacial acetic acid, determining the endpoint potentiometrically. Perform a blank determination, and make any necessary correction. Each mL of 0.1 N perchloric acid is equivalent to 12.82 mg of $C_3H_5KO_3$.

Chloride, page 471 Any turbidity produced by a quantity of the Solution containing the equivalent of 40 mg of potassium lactate does not exceed that shown in a control containing 20 µg of chloride ion (Cl).

Citrate, Oxalate, Phosphate, or Tartrate Dilute 5 mL with recently boiled and cooled water to 50 mL. To 4 mL of this solution, add 6 N ammonium hydroxide or 3 N hydrochloric acid, if necessary, to bring the pH to between 7.3 and 7.7. Add 1 mL of calcium chloride TS, and heat in a boiling water bath for 5 min: the solution remains clear.

Cyanide (*Caution*: Because of the extremely poisonous nature of potassium cyanide, conduct this test in a fume hood, and exercise great care to prevent skin contact and inhaling of particles or vapors of solutions of the material. Under no conditions pipet solutions by mouth.)

p-Phenylenediamine–Pyridine Mixed Reagent Dissolve 200 mg of p-phenylenediamine hydrochloride in 100 mL of water, warming to aid dissolution. Cool, allow the solids to settle, and use the supernatant liquid to make the mixed reagent. Dissolve 128 mL of pyridine in 365 mL of water, add 10 mL of hydrochloric acid, and mix. To prepare the mixed reagent, mix 30 mL of the p-phenylenediamine solution with all of the pyridine solution, and allow to stand for 24 h before using. The mixed reagent is stable for about 3 weeks when stored in an amber bottle.

Sample Solution Transfer an accurately weighed quantity of the Solution, equivalent to 20.0 g of potassium lactate, into a 100-mL volumetric flask, dilute to volume with water, and mix.

Cyanide Standard Solution Dissolve 100 mg of potassium cyanide, accurately weighed, in 10 mL of 0.1 N sodium hydroxide in a 100-mL volumetric flask, dilute to volume with 0.1 N sodium hydroxide, and mix. Transfer a 10-mL aliquot into a 1000-mL volumetric flask, dilute to volume with 0.1 N sodium hydroxide, and mix. Each mL of this solution contains 10 µg of cyanide.

Procedure Pipet a 10-mL aliquot of the *Sample Solution* into a 50-mL beaker. Into a second 50-mL beaker, pipet 0.10 mL of the *Cyanide Standard Solution*, and add 10 mL of water. Place the beakers in an ice bath, and adjust the pH to between 9 and 10 with 20% sodium hydroxide, stirring slowly and adding the reagent slowly to avoid overheating. Allow the solutions to stand for 3 min, and then slowly add 10% phosphoric acid to a pH between 5 and 6, measured with a pH meter. Transfer the solutions into 100-mL separators containing 25 mL of cold water, and rinse the beakers and pH meter electrodes with a few mL of cold water, collecting the washings in the respective separator. Add 2 mL of bromine TS, stopper, and mix. Add 2 mL of 2% sodium arsenite solution, stopper, and mix. Add 10 mL of n-butanol to the clear solutions, stopper, and mix. Finally, add 5 mL of *p-Phenylenediamine–Pyridine Mixed Reagent*, mix, and allow to stand for 15 min. Remove and discard the aqueous phases, and filter the alcohol phases into 1-cm cells. The absorbance of the solution from the *Sample Solution*, determined at 480 nm with a suitable spectrophotometer, is not greater than that from the *Cyanide Standard Solution*.

Heavy Metals Dilute a quantity of the Solution, equivalent to 2.0 g of potassium lactate, to 25 mL with water. This solution meets the requirements of the *Heavy Metals Test*, page 512, using 20 µg of lead ion (Pb) in the control (*Solution A*).

Lead Dilute a quantity of the Solution, equivalent to 2.0 g of potassium lactate, to 25 mL with water. This solution meets the requirements of the *Lead Limit Test*, page 518, using 10 µg of lead ion (Pb) in the control.

Methanol and Methyl Esters

Potassium Permanganate and Phosphoric Acid Solution Dissolve 3 g of potassium permanganate in a mixture of 15 mL of phosphoric acid and 70 mL of water. Dilute with water to 100 mL.

Oxalic Acid and Sulfuric Acid Solution Cautiously add 50 mL of sulfuric acid to 50 mL of water, mix, cool, add 5 g of oxalic acid, and mix to dissolve.

Standard Preparation Prepare a solution containing 10.0 mg of methanol in a 100-mL volumetric flask, dilute to volume with dilute alcohol (1 in 10), and mix.

Test Preparation Place 40.0 g of the Solution in a glass-stoppered, round-bottom flask, add 10 mL of water, and cautiously add 30 mL of 5 N potassium hydroxide. Connect a condenser to the flask, and steam-distill, collecting the distillate in a suitable 100-mL graduated vessel containing 10 mL of alcohol. Continue the distillation until the volume in the receiver reaches approximately 95 mL, and dilute the distillate with water to 100.0 mL.

Procedure Transfer 10.0 mL each of the *Standard Preparation* and the *Test Preparation* to separate 25-mL volumetric flasks. To each, add 5.0 mL of *Potassium Permanganate and Phosphoric Acid Solution*, and mix. After 15 min, add 2.0 mL of *Oxalic Acid and Sulfuric Acid Solution* to each, stir with a glass rod until the solution is colorless, add 5.0 mL of fuchsin–sulfurous acid TS (prepared as directed on page 173 of THIS SUPPLEMENT), dilute with water to volume, and mix. After 2 h, concomitantly determine the absorbances of both solutions in 1-cm cells at the wavelength of maximum absorbance at about 575 nm, with a suitable spectrophotometer and using water as the blank: the absorbance of the solution from the *Test Preparation* is not greater than that from the *Standard Preparation*.

pH Determine the pH of the Solution by the *Potentiometric Method*, page 531.

Sodium

Potassium Chloride Solution Dissolve 100 g of potassium chloride in water and dilute to 1000 mL.

Standard Solutions Transfer 127.1 mg of sodium chloride, previously dried at 105° for 2 h and accurately weighed, to a 500-mL volumetric flask, dilute with water to volume, and mix. Transfer 10.0 mL of this solution to a 100-mL volumetric flask, dilute with water to volume, and mix to obtain a *Stock Solution* containing 10 µg of sodium per mL. Into separate 100-mL volumetric flasks, pipet 1-, 2-, 5-, and 10-mL aliquots of the *Stock Solution*; add 1.0 mL of *Potassium Chloride Solution* followed by 1.0 mL of nitric acid; dilute with water to volume; and mix to obtain *Standard Solutions* containing 0.1, 0.2, 0.5, and 1.0 µg of sodium per mL, respectively.

Test Solution Transfer an accurately weighed quantity of the Solution equivalent to about 4 g of potassium lactate to a 50-mL volumetric flask, dilute to volume with water, and mix. Pipet 1 mL of this solution into a 100-mL volumetric flask, add 1.0 mL of *Potassium Chloride Solution* followed by 1.0 mL of nitric acid, dilute with water to volume, and mix.

Blank Solution Transfer 1.0 mL of *Potassium Chloride Solution* to a 100-mL volumetric flask, add 1.0 mL of nitric acid, dilute with water to volume, and mix.

Procedure Concomitantly determine the absorbances of the *Standard Solutions* and the *Test Solution* at the sodium emission line of 589 nm with a suitable atomic absorption spectrophotometer equipped with a sodium hollow-cathode lamp and an oxidizing air–acetylene flame, using the *Blank Solution* to zero the instrument. Plot the absorbances of the *Standard Solutions* versus concentration, in µg/mL, of sodium, and draw the straight line that best fits the plotted points. From the graph so obtained, determine the concentration C, in µg/mL, of sodium in the *Test Solution*. Calculate the percentage of sodium in the portion of potassium lactate taken by the formula:

$$CD/10{,}000W,$$

in which W is the quantity, in g, of potassium lactate taken to prepare the *Test Solution*, and D is the dilution factor for the *Test Solution*.

Sugars To 10 mL of hot alkaline cupric tartrate TS add 5 drops of Potassium Lactate Solution: no red precipitate is formed.

Sulfate, page 471 Any turbidity produced by a quantity of the Solution containing the equivalent of 4.0 g of potassium lactate does not exceed that shown in a control containing 200 µg of sulfate ion (SO_4).

Functional Use in Foods Emulsifier; flavor enhancer; flavoring agent or adjuvant; humectant; pH control agent.

Packaging and Storage Store in tight containers.

Potassium Sorbate, page 252

Change the *Structural Formula* to read:

$$CH_3CH{=}CHCH{=}CHCOOK$$

Change the second sentence in the *Description* to read:

It decomposes at about 270°.

Insert the following *Requirement* for *Lead*:

Lead Not more than 5 mg/kg.

Change the *Requirement* entitled *Loss on Drying* to read:

Loss on Drying Not more than 1.0%.

Insert the following *Test* for *Lead*:

Lead Prepare and test a 2-g sample as directed under the *Lead Limit Test*, page 518, using 10 µg of lead ion (Pb) in the control.

L-Proline, page 254

Change the *Requirements* entitled *Assay* and *Arsenic* to read:

Assay Not less than 98.5% and not more than 101.5% $C_5H_9NO_2$, calculated on the dried basis.

Arsenic (as As) Not more than 1.5 mg/kg.

Change the *Test* entitled *Arsenic* to read:

Arsenic   A *Sample Solution* prepared as directed for organic compounds meets the requirements of the *Arsenic Test*, page 464, using 1.5 mL of the *Standard Arsenic Solution* in the control (1.5 μg As).

Insert the following new monograph to precede the monograph entitled *Propionic Acid*, page 254:

Propane

C_3H_8 Mol wt 44.10

DESCRIPTION

A colorless, odorless, flammable gas (boiling temperature is about –42°). One hundred volumes of water dissolves 6.5 volumes at 17.8° and a pressure of 753 mm of mercury; 100 volumes of anhydrous alcohol dissolves 790 volumes at 16.6° and a pressure of 754 mm of mercury; 100 volumes of ether dissolves 926 volumes at 16.6° and a pressure of 757 mm of mercury; 100 volumes of chloroform dissolves 1299 volumes at 21.6° and a pressure of 757 mm of mercury. Vapor pressure at 21° is about 10,290 mm of mercury (108 psi).

REQUIREMENTS

Caution: Propane is highly flammable and explosive. Observe precautions and perform sampling and analytical operations in a well-ventilated fume hood.

Identification

A. The infrared absorption spectrum exhibits maxima, among others, at about the following wavelengths, in μm: 3.4 (vs), 6.8 (s), and 7.2 (m).
B. The vapor pressure of a test specimen, obtained as directed in the *Sampling Procedure*, and determined at 21° by means of a suitable pressure gauge, is between 820 and 875 kPa absolute (119 and 127 psia, respectively).

Assay   Not less than 98.0% propane (C_3H_8).
Acidity of Residue   Passes test.
High-Boiling Residue   Not more than 5 mg/kg.
Sulfur Compounds   Passes test.
Water   Not more than 10 mg/kg.

TESTS

Sampling Procedure, Assay, Acidity of Residue, High-Boiling Residue, Sulfur Compounds, and **Water**   Proceed as directed for these tests under *Butane*, page 101 of THIS SUPPLEMENT, except substitute Propane wherever Butane is specified in the text.

Functional Use in Foods   Propellant; aerating agent.
Packaging and Storage   Store in tight cylinders protected from excessive heat.

Insert the following new monographs to precede the monograph entitled *Riboflavin*, page 262:

Rapeseed Oil, Fully Hydrogenated

Fully Hydrogenated Rapeseed Oil

DESCRIPTION

A white, waxy, odorless solid that is a mixture of triglycerides. The saturated fatty acids are found in the same proportions that result from the full hydrogenation of fatty acids occurring in natural high erucic acid rapeseed oil. The rapeseed oil is obtained from *Brassica napus* and *Brassica campestris* of the family *Cruciferae*. It is made by hydrogenating high erucic acid rapeseed oil in the presence of a nickel catalyst at temperatures not exceeding 245°.

REQUIREMENTS

Identification

Fully hydrogenated rapeseed oil exhibits the following composition profile of fatty acids as determined under *Fatty Acid Composition*, page 82 of the Second Supplement.

Fatty Acid:	14:0	16:0	18:0	18:1	18:2
Weight % (Range):	<1.0	3–5	38–42	1.0	<1.0
Fatty Acid:	20:0	20:1	22:0	22:1	24:0
Weight % (Range):	8–10	<1.0	42–50	<1.0	1.0–2.0

Acid Value   Not more than 6.
Arsenic (as As)   Not more than 0.5 mg/kg.
Cold Test   Passes test.
Color (Lovibond)   Not more than 1.5 red/15 yellow.
Erucic Acid   Not more than 1.0%.
Free Fatty Acids (as oleic acid)   Not more than 2.0%.
Heavy Metals (as Pb)   Not more than 5 mg/kg.
Iodine Value   Not more than 4.
Lead   Not more than 0.1 mg/kg.
Peroxide Value   Not more than 2.0 meq/kg.
Residue on Ignition   Not more than 0.5%.
Unsaponifiable Matter   Not more than 1.5%.
Water   Not more than 0.05%.

ADDITIONAL REQUIREMENTS

The following specification should conform to the representations of the vendor: *1-Monoglyceride Content*.

Labeling Rapeseed oil products that have been fully hydrogenated should be labeled as fully hydrogenated rapeseed oil. Label to indicate 1-monoglyceride content.

TESTS

1-Monoglyceride Content Determine as directed in the general method, page 506.
Acid Value Determine as directed under *Method II* in the general procedure, page 504.
Arsenic A *Sample Solution* containing 2.0 g of the sample prepared as directed for organic compounds meets the requirements of the *Arsenic Test*, page 464, using 1 mL of the *Standard Arsenic Solution* in the control (1 µg As).
Cold Test Proceed as directed under *Cold Test*, page 82 of the Second Supplement.
Color Proceed as directed under *Color*, page 82 of the Second Supplement. Use a 13.34-cm cell.
Erucic Acid Determine as part of *Fatty Acid Composition*, page 82 of the Second Supplement.
Free Fatty Acids Proceed as directed under *Free Fatty Acids*, page 504, using the following equivalence factor (e) in the formula given in the procedure:

$$\text{Free fatty acids as oleic acid, } e = 28.2$$

Heavy Metals Prepare and test a 2-g sample as directed in *Method II* under the *Heavy Metals Test*, page 513, using 10 µg of lead ion (Pb) in the control (*Solution A*).
Iodine Value Proceed as directed under *Wijs Method*, page 505.
Lead Determine as directed under *Method II* in the *Atomic Absorption Spectrophotometric Graphite Furnace Method* under the *Lead Limit Test*, page 169 of THIS SUPPLEMENT, using a 10-g sample.
Peroxide Value Proceed as directed under *Peroxide Value*, page 148 of the monograph for *Hydroxylated Lecithin*. However, after the addition of saturated potassium iodide and mixing, mix the solution for only 1 min and begin the titration immediately instead of allowing the solution to stand for 10 min.
Residue on Ignition Ignite 5 g as directed in the general method, page 533.
Unsaponifiable Matter Proceed as directed under *Unsaponifiable Matter*, page 509.
Water Proceed as directed under *Water Determination* by the *Karl Fischer Titrimetric Method*, page 553. However, in place of 35–40 mL of methanol, use 50 mL of a 1:1 chloroform–methanol mixture to dissolve the sample.

Functional Use in Foods Coating agent; emulsifying agent; stabilizer; thickener; formulation aid; texturizer.
Packaging and Storage Store in well-closed containers.

Rapeseed Oil, Superglycerinated

Superglycerinated Fully Hydrogenated Rapeseed Oil

DESCRIPTION

A white solid that is a mixture of mono-, di-, and triglycerides with triglycerides as a minor component. The saturated fatty acids are found in the same proportions that result from the full hydrogenation of fatty acids occurring in natural high erucic acid rapeseed oil. The rapeseed oil is typically obtained by *n*-hexane extraction from *Brassica napus* and *Brassica campestris* of the family *Cruciferae*. It is made by adding excess glycerin to fully hydrogenated rapeseed oil and heating, in the presence of sodium hydroxide catalyst, to about 165° under partial vacuum and steam sparging agitation.

REQUIREMENTS

Identification

Superglycerinated rapeseed oil exhibits the same fatty acid composition as fully hydrogenated rapeseed oil. It exhibits the following composition profile of fatty acids as determined under *Fatty Acid Composition*, page 82 of the Second Supplement.

Fatty Acid:	14:0	16:0	18:0	18:1	18:2
Weight % (Range):	<1.0	3–5	38–42	1.0	<1.0
Fatty Acid:	20:0	20:1	22:0	22:1	24:0
Weight % (Range):	8–10	<1.0	42–50	<1.0	1.0–2.0

Acid Value Not more than 6.
Arsenic (as As) Not more than 0.5 mg/kg.
Cold Test Passes test.
Color (Lovibond) Not more than 1.5 red/15 yellow.
Erucic Acid Not more than 1.0%.
Free Fatty Acids (as oleic acid) Not more than 2.0%.
Free Glycerin Not more than 1%.
Heavy Metals Not more than 5 mg/kg.
Iodine Value Not more than 4.
Lead Not more than 0.1 mg/kg.
Peroxide Value Not more than 2.0 meq/kg.
Residue on Ignition Not more than 0.5%.
Unsaponifiable Matter Not more than 1.5%.
Water Not more than 0.05%.

ADDITIONAL REQUIREMENTS

The following specifications should conform to the representations of the vendor: *1-Monoglyceride Content* and *Hydroxyl Value*.

Labeling Rapeseed oil products that have added glycerin (glycerol) and are fully hydrogenated must be labeled as fully hydrogenated and superglycerinated rapeseed oil. Label to indicate 1-monoglyceride content and hydroxyl value.

TESTS

1-Monoglyceride Content Determine as directed in the general method, page 506.
Acid Value Determine as directed under *Method II* in the general procedure, page 504.
Arsenic A *Sample Solution* containing 2.0 g of the sample prepared as directed for organic compounds meets the requirements of the *Arsenic Test*, page 464, using 1 mL of the *Standard Arsenic Solution* in the control (1 µg As).
Cold Test Proceed as directed under *Cold Test*, page 82 of the Second Supplement.
Color Proceed as directed under *Color*, page 82 of the Second Supplement. Use a 13.34-cm cell.
Erucic Acid Determine as part of *Fatty Acid Composition*, page 82 of the Second Supplement.
Free Fatty Acids Proceed as directed under *Free Fatty Acids*, page 504, using the following equivalence factor (*e*) in the formula given in the procedure:

Free fatty acids as oleic acid, *e* = 28.2

Free Glycerin Determine as directed in the general method, page 504.
Heavy Metals Prepare and test a 2-g sample as directed in *Method II* under the *Heavy Metals Test*, page 513, using 10 µg of lead ion (Pb) in the control (*Solution A*).
Hydroxyl Value Determine as directed under *Method II* in the general procedure, page 504.
Iodine Value Proceed as directed under the *Wijs Method*, page 505.
Lead Determine as directed under *Method II* in the *Atomic Absorption Spectrophotometric Graphite Furnace Method* under the *Lead Limit Test*, page 169 of THIS SUPPLEMENT, using a 10-g sample.
Peroxide Value Proceed as directed under *Peroxide Value*, page 148 of the monograph for *Hydroxylated Lecithin*. However, after the addition of saturated potassium iodide and mixing, mix the solution for only 1 min and begin the titration immediately instead of allowing the solution to stand for 10 min.
Residue on Ignition Ignite 5 g as directed in the general method, page 533.

Total Monoglycerides Determine as directed in the general method, page 506.
Unsaponifiable Matter Proceed as directed under *Unsaponifiable Matter*, page 509.
Water Proceed as directed under *Water Determination* using the *Karl Fischer Titrimetric Method*, page 553. However, in place of 35–40 mL of methanol, use 50 mL of a 1:1 chloroform–methanol mixture to dissolve the sample.

Packaging and Storage Store in well-closed containers.
Functional Use in Foods Coating agent; emulsifying agent; formulation aid; texturizer.

DL-Serine, page 269

Change the *Requirements* entitled *Assay* and *Arsenic* to read:

Assay Not less than 98.5% and not more than 101.5% $C_3H_7NO_3$, calculated on the dried basis.
Arsenic (as As) Not more than 1.5 mg/kg.

Replace the *Test* entitled *Assay* with the following:

Assay Dissolve about 200 mg, accurately weighed, in 3 mL of formic acid and 50 mL of glacial acetic acid. Titrate with 0.1 *N* perchloric acid in glacial acetic acid, determining the endpoint potentiometrically. Perform a blank determination (see page 2), and make any necessary correction. Each mL of 0.1 *N* perchloric acid is equivalent to 10.51 mg of $C_3H_7NO_3$.

Change the *Test* entitled *Arsenic* to read:

Arsenic A *Sample Solution* prepared as directed for organic compounds meets the requirements of the *Arsenic Test*, page 464, using 1.5 mL of the *Standard Arsenic Solution* in the control (1.5 µg As).

L-Serine, page 270

Change the *Requirements* entitled *Assay*, *Arsenic*, and *Specific Rotation* to read:

Assay Not less than 98.5% and not more than 101.5% $C_3H_7NO_3$, calculated on the dried basis.
Arsenic (as As) Not more than 1.5 mg/kg.
Specific Rotation $[\alpha]_D^{20°}$: Between +13.6° and +15.6°, calculated on the dried basis.

Replace the *Test* entitled *Assay* with the following:

Assay Dissolve about 200 mg, accurately weighed, in 3 mL of formic acid and 50 mL of glacial acetic acid. Titrate with 0.1 N perchloric acid in glacial acetic acid, determining the endpoint potentiometrically. Perform a blank determination (see page 2), and make any necessary correction. Each mL of 0.1 N perchloric acid is equivalent to 10.51 mg of $C_3H_7NO_3$.

Change the *Test* entitled *Arsenic* to read:

Arsenic A *Sample Solution* prepared as directed for organic compounds meets the requirements of the *Arsenic Test*, page 464, using 1.5 mL of the *Standard Arsenic Solution* in the control (1.5 μg As).

Sodium Aluminosilicate, page 274

Change the first sentence of the *Description* to read:

A series of hydrated sodium aluminum silicates having an $Na_2O/Al_2O_3/SiO_2$ mol ratio of approximately 1/1/13, respectively.

Change the *Requirements* entitled *Assay, Silicon Dioxide*; *Loss on Drying*; and *Loss on Ignition* to read:

Assay
  Silicon Dioxide Not less than 66.0% and not more than 76.0% SiO_2 after drying.
Loss on Drying Not more than 8.0%.
Loss in Ignition Between 8.0% and 11.0%.

Change the first and fourth sentences of the first paragraph of the *Test* entitled *Assay, Silicon Dioxide*, to read, respectively:

Transfer about 500 mg, previously dried at 105° for 2 h and accurately weighed, into a 250-mL beaker, wash the sides of the beaker with a few mL of water, and then add 30 mL of sulfuric acid and 15 mL of hydrochloric acid.

Wash the filter paper and precipitate with hot water until the filter paper is free of sulfuric acid.

Delete the last sentence of the last paragraph of the *Test* entitled *Assay, Silicon Dioxide*.

Replace the *Test* entitled *Assay, Aluminum Oxide*, with the following:

Transfer about 500 mg of the sample, previously dried at 105° for 2 h and accurately weighed, into a tared platinum dish, and moisten with 8 to 10 drops of water. Add 25 mL of 70% perchloric acid and 10 mL of hydrofluoric acid, and heat on a hot plate in a hood until dense, white fumes of perchloric acid appear. Cool, add 10 mL of hydrofluoric acid, and heat again to dense, white fumes. Cool, dissolve the residue in sufficient water, quantitatively transfer with the aid of additional water to a 250-mL volumetric flask, and dilute to volume. Retain this solution for sodium oxide analysis.

Transfer by pipet a 10.0-mL aliquot of this solution into a 100-mL volumetric flask, fill to volume with water, and mix.

Set a suitable atomic absorption spectrophotometer to a wavelength of 309.3 nm. Adjust the instrument to zero absorbance against water. Read the absorbance of four standard solutions containing 5, 10, 20, and 50 μg/mL of aluminum, in the form of the chloride, and plot the standard curve as absorbance versus concentration of aluminum.

Aspirate the 1 in 10 diluted sample solution into the spectrophotometer, read the absorbance in the same manner, and by reference to the standard curve, determine the concentration (C) of aluminum, in μg/mL, in the sample solution.

Calculate the quantity, in mg, of Al_2O_3 in the sample taken by the formula:

$$250C \times 10 \times 1.8895/1000.$$

Delete the first paragraph of the *Test* entitled *Assay, Sodium Oxide*.

Replace the first sentence of the last paragraph of the *Test* entitled *Assay, Sodium Oxide*, with the following:

Place a portion of the sample solution prepared for the aluminum oxide determination in the photometer, read the percent transmittance in the same manner, and by reference to the standard curve, determine the concentration (C) of sodium, in μg/mL, in the sample solution.

Replace the *Test* entitled *Assay, Correction for Sodium Sulfate Content*, with the following:

Transfer about 1 g of the sample, previously dried at 105° for 2 h and accurately weighed, into a tared platinum dish, and moisten with 8 to 10 drops of water. Add 25 mL of 70% perchloric acid and 10 mL of hydrofluoric acid, and heat on a hot plate in a hood until dense, white fumes of perchloric acid appear. Add 10 mL of hydrofluoric acid, and heat again to dense, white fumes. Quantitatively transfer the solution to a 400-mL beaker, add 200 mL of water, and heat to boiling. Gradually add, in small portions at a time and while stirring constantly, an excess of hot barium chloride TS (about 10 mL), and heat the mixture on a steam bath for 1 h. Collect the precipitate on a filter, wash until free from chloride, dry, ignite, and weigh. The weight, in g, of the barium sulfate so obtained, multiplied by 0.6086, indicates its equivalent of Na_2SO_4 (C'). Calculate the correction factor (F) by the formula:

$$0.437(C' \times w/W),$$

in which w is the weight, in mg, of the sample taken for the sodium oxide determination, and W is the weight, in mg, of the sample taken for the sodium sulfate determination.

Insert the following new monograph to precede the monograph entitled *Sodium Lauryl Sulfate*, page 289:

Sodium Lactate Solution

Propanoic Acid, 2-Hydroxy-, Monosodium Salt

$C_3H_5NaO_3$ Formula wt 112.06

DESCRIPTION

Sodium lactate solution is a clear, colorless or practically colorless, slightly viscous liquid, odorless or having a slight, not unpleasant odor. It is miscible with water, and it is normally available in a concentration range of 60% to about 80% $C_3H_5NaO_3$, by weight.

REQUIREMENTS

Identification

It gives positive tests for *Sodium*, page 517, and for *Lactate*, page 517.

Assay Not less than 50.0%, by weight, and not less than 98.0% and not more than 102.0%, by weight, of the labeled amount of $C_3H_5NaO_3$.
Arsenic (as As) Not more than 1 mg/kg.
Chloride Not more than 0.05%.
Citrate, Oxalate, Phosphate, or Tartrate Passes test.
Cyanide Not more than 0.5 mg/kg.
Heavy Metals (as Pb) Not more than 10 mg/kg.
Lead Not more than 5 mg/kg.
Methanol and Methyl Esters Not more than 0.025%.
pH Between 5.0 and 9.0.
Sugars Passes test.
Sulfate Not more than 0.005%.

ADDITIONAL REQUIREMENTS

Labeling Label to indicate its content, by weight, of sodium lactate, $C_3H_5NaO_3$.

TESTS

Assay Weigh accurately into a suitable flask a volume of Sodium Lactate Solution, equivalent to about 300 mg of sodium lactate. Add 60 mL of a 1 in 5 mixture of acetic anhydride in glacial acetic acid, mix, and allow to stand for 20 min. Titrate with 0.1 N perchloric acid in glacial acetic acid, determining the endpoint potentiometrically. Perform a blank determination, and make any necessary correction. Each mL of 0.1 N perchloric acid is equivalent to 11.21 mg of $C_3H_5NaO_3$.

Arsenic A *Sample Solution* containing the equivalent of 1.0 g of sodium lactate meets the requirements of the *Arsenic Test*, page 464, using 1 mL of the *Standard Arsenic Solution* in the control (1 μg As).

Chloride, page 471 Any turbidity produced by a quantity of the solution containing the equivalent of 40 mg of sodium lactate does not exceed that shown in a control containing 20 μg of chloride ion (Cl).

Citrate, Oxalate, Phosphate, or Tartrate Dilute 5 mL with recently boiled and cooled water to 50 mL. To 4 mL of this solution add 6 N ammonium hydroxide or 3 N hydrochloric acid, if necessary, to bring the pH to between 7.3 and 7.7. Add 1 mL of calcium chloride TS, and heat in a boiling water bath for 5 min: the solution remains clear.

Cyanide (*Caution*: Because of the extremely poisonous nature of potassium cyanide, conduct this test in a fume hood, and exercise great care to prevent skin contact and inhaling particles or vapors of solutions of the material. Under no conditions pipet solutions by mouth.)

p-Phenylenediamine–Pyridine Mixed Reagent Dissolve 200 mg of *p*-phenylenediamine hydrochloride in 100 mL of water, warming to aid dissolution. Cool, allow the solids to settle, and use the supernatant liquid to make the mixed reagent. Dissolve 128 mL of pyridine in 365 mL of water, add 10 mL of hydrochloric acid, and mix. To prepare the mixed reagent, mix 30 mL of the *p*-phenylenediamine solution with all of the pyridine solution and allow to stand for 24 h before using. The mixed reagent is stable for about 3 weeks when stored in an amber bottle.

Sample Solution Transfer an accurately weighed quantity of the Solution, equivalent to 20.0 g of sodium lactate, into a 100-mL volumetric flask, dilute to volume with water, and mix.

Cyanide Standard Solution Dissolve 100 mg of potassium cyanide, accurately weighed, in 10 mL of 0.1 N sodium hydroxide in a 100 mL volumetric flask, dilute to volume with 0.1 N sodium hydroxide, and mix. Transfer a 10-mL aliquot into a 1000-mL volumetric flask, dilute to volume with 0.1 N sodium hydroxide, and mix. Each mL of this solution contains 10 μg of cyanide.

Procedure Pipet a 10-mL aliquot of the *Sample Solution* into a 50-mL beaker. Into a second 50-mL beaker, pipet 0.1 mL of the *Cyanide Standard Solution*, and add 10 mL of water. Place the beakers in an ice bath, and adjust the pH to between 9 and 10 with 20% sodium hydroxide, stirring slowly and adding the reagent slowly to avoid overheating. Allow the solutions to stand for 3 min, and then slowly add 10% phosphoric acid to a pH between 5 and 6, measured with a pH meter.

Transfer the solutions into 100-mL separators containing 25 mL of cold water, and rinse the beakers and pH meter electrodes with a few mL of cold water, collecting the washings in the respective separator. Add 2 mL of bromine TS, stopper, and mix. Add 2 mL of 2% sodium arsenite solution, stopper, and mix. To the clear solutions add 10 mL of *n*-butanol, stopper, and mix. Finally, add 5 mL of *p-Phenylenediamine–Pyridine Mixed Reagent*, mix, and allow to stand for 15 min. Remove and discard the aqueous phases, and filter the alcohol phases into 1-cm cells. The absorbance of the solution from the *Sample Solution*, determined at 480 nm with a suitable spectrophotometer, is no greater than that from the *Cyanide Standard Solution*.

Heavy Metals Dilute a quantity of the Solution, equivalent to 2.0 g of sodium lactate, to 25 mL with water. This solution meets the requirements of the *Heavy Metals Test*, page 512, using 20 μg of lead ion (Pb) in the control (*Solution A*).

Lead Dilute a quantity of the Solution, equivalent to 2.0 g of sodium lactate, to 25 mL with water. This solution meets the requirements of the *Lead Limit Test*, page 518, using 10 μg of lead ion (Pb) in the control.

Methanol and Methyl Esters

Potassium Permanganate and Phosphoric Acid Solution Dissolve 3 g of potassium permanganate in a mixture of 15 mL of phosphoric acid and 70 mL of water. Dilute with water to 100 mL.

Oxalic Acid and Sulfuric Acid Solution Cautiously add 50 mL of sulfuric acid to 50 mL of water, mix, cool, add 5 g of oxalic acid, and mix to dissolve.

Standard Preparation Prepare a solution containing 10.0 mg of methanol in 100 mL of dilute alcohol (1 in 10).

Test Preparation Place 40.0 g of the Solution in a glass-stoppered, round-bottom flask, add 10 mL of water, and add cautiously 30 mL of 5 *N* potassium hydroxide. Connect a condenser to the flask, and steam-distill, collecting the distillate in a suitable 100-mL graduated vessel containing 10 mL of alcohol. Continue the distillation until the volume in the receiver reaches approximately 95 mL, and dilute the distillate with water to 100.0 mL.

Procedure Transfer 10.0 mL each of the *Standard Preparation* and the *Test Preparation* to 25-mL volumetric flasks. To each add 5.0 mL of *Potassium Permanganate and Phosphoric Acid Solution*, and mix. After 15 min, to each add 2.0 mL of *Oxalic Acid and Sulfuric Acid Solution*, stir with a glass rod until the solution is colorless, add 5.0 mL of fuchsin–sulfurous acid TS (prepared as directed on page 173 of THIS SUPPLEMENT), dilute with water to volume, and mix. After 2 h, concomitantly determine the absorbances of both solutions in 1-cm cells at the wavelength of maximum absorbance at about 575 nm, with a suitable spectrophotometer, using water as the blank: the absorbance of the solution from the *Test Preparation* is not greater than that from the *Standard Preparation*.

pH Determine the pH of the Solution by the *Potentiometric Method*, page 531.

Sugars To 10 mL of hot alkaline cupric tartrate TS add 5 drops of Sodium Lactate Solution: no red precipitate is formed.

Sulfate, page 471 Any turbidity produced by a quantity of the Solution containing the equivalent of 4.0 g of sodium lactate does not exceed that shown in a control containing 200 μg of sulfate ion (SO_4).

Functional Use in Foods Emulsifier; flavor enhancer; flavoring agent or adjuvant; humectant; pH control agent.

Packaging and Storage Store in tight containers.

Insert the following new monograph to precede the monograph entitled *Sodium Metabisulfite*, page 289:

Sodium Magnesium Aluminosilicate

DESCRIPTION

A series of hydrated sodium magnesium aluminosilicates having $Na_2O:MgO:Al_2O_3:SiO_2$ molar ratios of approximately 2:1:2:24, respectively. They are synthetic, amorphous, food-grade coprecipitates that are fine, white powders or beads with a specific gravity of about 2. They are odorless, tasteless, and insoluble in water, in alcohol, and in other organic solvents, but are partially soluble in strongly acidic and alkaline solutions.

REQUIREMENTS

Identification

A. Mix 500 mg of the sample with 2.5 g of anhydrous potassium carbonate, and heat the mixture in a platinum or nickel crucible until it melts completely. Cool, add 5 mL of water, and allow to stand for 3 min. Heat the bottom of the crucible gently, detach the melt, and transfer it to a beaker with the aid of about 50 mL of water. Gradually add hydrochloric acid until no effervescence is observed, add 10 mL more of the acid, and evaporate to dryness on a steam bath. Cool, add 20 mL of water, boil, and filter through ash-free paper. An insoluble residue of silica remains. (NOTE: Retain the filtrate for *Identification Test B*.) Transfer the gelatinous residue to a platinum dish, and cautiously add 5 mL of hydrofluoric acid. The precipitate dissolves. (If it does not dissolve, repeat the treatment with hydrofluoric acid.) Heat, and introduce into the resulting vapors a glass stirring rod with a drop of water on the tip. The drop becomes turbid.

B. Portions of the filtrate obtained in *Identification Test A* give positive tests for *Aluminum*, page 515, and for *Sodium* and *Magnesium*, page 517.

Assay

Silicon Dioxide   Not less than 65.0% and not more than 75.0% SiO_2 after drying.

Aluminum Oxide   Not less than 9.0% and not more than 13.0% Al_2O_3 after drying.

Magnesium Oxide   Not less than 1.0% and not more than 3.0% MgO after drying.

Sodium Oxide   Not less than 3.0% and not more than 9.0% Na_2O after drying.

Arsenic (as As)   Not more than 3 mg/kg.
Heavy Metals (as Pb)   Not more than 10 mg/kg.
Loss on Drying   Not more than 8.0%.
Loss on Ignition   Between 8.0% and 11.0% after drying.
pH   Between 6.5 and 11.0.
Soluble Salt (as Na_2SO_4)   Not more than 7.5%.

TESTS

Assay

Silicon Dioxide   Transfer about 500 mg, previously dried at 105° for 2 h and accurately weighed, into a 250-mL beaker, wash the sides of the beaker with a few mL of water, and then add 30 mL of sulfuric acid and 15 mL of hydrochloric acid. Heat on a hot plate in a hood until dense, white fumes are evolved, cool, add 15 mL of hydrochloric acid, and heat again to dense, white fumes. Cool, add 70 mL of water, and filter through Whatman No. 40 (or an equivalent) filter paper. Wash the filter paper and precipitate thoroughly with hot water to remove the sulfuric acid residue.

Transfer the filter paper and precipitate into a tared platinum crucible, char, and ignite at 900° to constant weight. Moisten the residue with a few drops of water, add 15 mL of hydrofluoric acid and 8 drops of sulfuric acid, and heat on a hot plate in a hood until white fumes of sulfur trioxide are evolved. Cool; add 5 mL of water, 10 mL of hydrofluoric acid, and 3 drops of sulfuric acid; and evaporate to dryness on the hot plate. Heat cautiously over an open flame until sulfur trioxide fumes have ceased, and ignite at 900° to constant weight. The weight loss after the addition of hydrofluoric acid represents the weight of SiO_2 in the sample taken.

Aluminum Oxide   Transfer about 500 mg of the sample, previously dried at 105° for 2 h and accurately weighed, into a tared platinum dish, and moisten with 8 to 10 drops of water. Add 25 mL of 70% perchloric acid and 10 mL of hydrofluoric acid, and heat on a hot plate in a hood until dense, white fumes of perchloric acid appear. Cool, add 10 mL of hydrofluoric acid, and heat again to dense, white fumes. Cool, dissolve the residue in sufficient water, quantitatively transfer with the aid of additional water to a 250-mL volumetric flask, dilute to volume with water, and mix. Retain this solution for analysis under *Magnesium Oxide* and *Sodium Oxide*.

Transfer by pipet a 10.0-mL aliquot of this solution into a 100-mL volumetric flask, dilute to volume with water, and mix.

Set a suitable atomic absorption spectrophotometer to a wavelength of 309.3 nm. Adjust the instrument to zero absorbance against water. Read the absorbance of four standard solutions containing 5, 10, 20, and 50 µg/mL of aluminum, in the form of the chloride, and plot the standard curve as absorbance versus concentration of aluminum.

Aspirate a 1 in 10 diluted sample solution into the spectrophotometer, read the absorbance in the same manner, and by reference to the standard curve, determine the concentration (C) of aluminum, in µg/mL, in the sample solution.

Calculate the quantity, in mg, of Al_2O_3 in the sample taken by the formula:

$$250C \times 10 \times 1.8895/1000.$$

Magnesium Oxide   Set a suitable atomic absorption spectrophotometer to a wavelength of 285.2 nm. Adjust the instrument to zero absorbance against water. Read the absorbance of four standard solutions containing 5, 10, 25, and 50 µg/mL of magnesium, in the form of the chloride, and plot the standard curve as absorbance versus concentration of magnesium.

Aspirate the sample solution prepared for the aluminum determination into the spectrophotometer, read the absorbance in the same manner, and by reference to the standard curve, determine the concentration (C) of magnesium, in µg/mL, in the sample solution. Calculate the quantity, in mg, of MgO in the sample taken by the formula:

$$250C \times 1.6579/1000.$$

Sodium Oxide   Set a suitable flame photometer to a wavelength of 589 nm. Adjust the instrument to zero transmittance against water, and then adjust it to 100.0% transmittance with a standard solution containing 200 µg/mL of sodium, in the form of the chloride. Read the percent transmittance of three other standard solutions containing 50, 100, and 150 µg/mL each of sodium, and plot the standard curve as percent transmittance versus concentration of sodium.

Aspirate the sample solution prepared for the aluminum determination into the photometer, read the percent transmittance in the same manner, and by reference to the standard curve, determine the concentration (C) of sodium, in µg/mL, in the sample solution. Calculate the quantity, in mg, of Na_2O in the sample taken by the formula:

$$(250C \times 1.348/1000) - F,$$

in which F, as determined below, is the quantity of sodium oxide equivalent to any sodium sulfate present in the sample.

Correction for Sodium Sulfate Content   Transfer about 1 g of the sample, previously dried at 105° for 2 h and accurately weighed, into a tared platinum dish, and moisten with 8 to 10 drops of water. Add 25 mL of 70% perchloric acid and 10 mL of hydrofluoric acid, and heat on a hot plate in a hood until dense, white fumes of perchloric acid appear. Add 10 mL of hydrofluoric acid, and heat again to dense, white fumes. Quantitatively transfer the solution to a 400-mL beaker, add 200 mL of water, and heat to boiling. Gradually add, in small portions at a time and while stirring constantly, an excess of hot barium chloride TS (about 10 mL), and heat the mixture on a steam bath for 1 h. Collect the precipitate on a filter, wash until free from chloride, dry, ignite, and weigh. The weight, in mg, of the barium sulfate so obtained, multiplied by 0.6086, indicates its equivalent of Na_2SO_4 (C'). Calculate the correction factor (F) by the formula:

$$0.437(C' \times w/W),$$

in which w is the weight, in mg, of the sample taken for the sodium oxide determination, and W is the weight, in mg, of the sample taken for the sodium sulfate determination.

Sample Solution for the Determination of Arsenic and Heavy Metals   Transfer 10.0 g of the sample into a 250-mL beaker, add 50 mL of 0.5 N hydrochloric acid, cover with a watch glass, and heat slowly to boiling. Boil gently for 15 min, cool, and let the undissolved material settle. Decant the supernatant liquid through Whatman No. 4 (or an equivalent) filter paper into a 100-mL volumetric flask, retaining as much as possible of the insoluble material in the beaker. Wash the slurry and beaker with three 10-mL portions of hot water, decanting each washing through the filter into the flask. Finally, wash the filter paper with 15 mL of hot water, cool the filtrate to room temperature, dilute to volume with water, and mix.

Arsenic   A 10-mL portion of the *Sample Solution* meets the requirements of the *Arsenic Test*, page 464.

Heavy Metals   A 20-mL portion of the *Sample Solution* meets the requirements of the *Heavy Metals Test*, page 512, using 20 μg of lead ion (Pb) in the control (*Solution A*).

Loss on Drying, page 518   Dry at 105° for 2 h.

Loss on Ignition   Transfer about 5 g, previously dried at 105° for 2 h and accurately weighed, into a suitable tared crucible, and ignite at 900° to constant weight.

pH   Determine using a 1 in 5 slurry by the *Potentiometric Method*, page 531.

Soluble Salt   Calculate the percent sodium sulfate from the weight of barium sulfate obtained in the *Correction for Sodium Sulfate Content* in the *Assay*, by the formula:

$$N \times 60.86/W,$$

in which N is the weight, in mg, of barium sulfate, and W is the weight, in mg, of the sample taken for sodium sulfate determination.

Functional Use in Foods   Anticaking agent.

Packaging and Storage   Store in well-closed containers.

Sodium Stearyl Fumarate, page 301

Replace the *Identification* with the following:

The infrared absorption spectrum of a 1 in 300 potassium bromide dispersion of the sample exhibits maxima only at the same wavelengths as those of a similar preparation of USP Sodium Stearyl Fumarate Reference Standard.

Change the *Test* entitled *Sodium Stearyl Maleate and Stearyl Alcohol, Standard Solution A*, to read:

Weigh accurately 10 mg of USP Sodium Stearyl Maleate Reference Standard into a 100-mL volumetric flask, dilute to volume with 10% acetic acid in chloroform, and shake well.

Change the *Test* entitled *Sodium Stearyl Maleate and Stearyl Alcohol, Standard Solution B*, to read:

Weigh accurately 20 mg of stearyl alcohol (Aldrich or equivalent) into a 100-mL volumetric flask, dilute to volume with 10% acetic acid in chloroform, and shake well.

Sorbitol, page 308

Change the first sentence of the *Identification* to read:

Dissolve about 5 g of the sample in 6 mL of water; add 7 mL of methanol, 1 mL of benzaldehyde, and 1 mL of hydrochloric acid; and shake in a mechanical shaker until crystals appear.

Change the *Requirement* entitled *Assay* to read:

Assay   Not less than 91.0% and not more than 100.5% sorbitol ($C_6H_{14}O_6$), calculated on the anhydrous basis.

Delete the *Requirement* entitled *Loss on Drying*.

Insert the following under *Requirements*:

Water   Not more than 1.0%.

Replace the *Test* entitled *Assay* with the following:

Assay
Mobile Phase Use degassed water.
Standard Solution Dissolve an accurately weighed quantity of USP Sorbitol Reference Standard in water to obtain a solution with a known concentration of about 4.8 mg/mL.
Resolution Solution Dissolve USP Mannitol Reference Standard and USP Sorbitol Reference Standard in water to obtain a solution with concentrations of about 4.8 mg/mL of each.
Assay Solution Transfer about 240 mg of the sample, accurately weighed, to a 50-mL volumetric flask, dilute with water to volume, and mix.
Chromatographic System Use a high-pressure liquid chromatograph equipped with a refractive index detector that is maintained at constant temperature and a 7.8-mm × 30-cm column packed with a strong cation-exchange resin consisting of sulfonated, cross-linked styrene–divinylbenzene copolymer in the calcium form that is about 9 μm in diameter (Sugar Pak or equivalent) and that is maintained at 30° ± 2°. The flow rate is about 0.2 mL/min. Chromatograph the *Standard Solution* and record the peak responses. The relative standard deviation for three replicate injections is not more than 2.0%. Chromatograph the *Resolution Solution*: the resolution, R, between the sorbitol and mannitol peaks is not less than 2.0.
Procedure Separately inject equal volumes, about 20 μL each, of the *Assay Solution* and the *Standard Solution* into the chromatograph, and record the peak responses for the major peaks. Calculate the quantity, in mg, of $C_6H_{14}O_6$ in the Sorbitol taken by the formula:

$$50C(r_u/r_s),$$

in which C is the concentration, in mg/mL, of USP Sorbitol Reference Standard in the *Standard Solution*, and r_u and r_s are the peak responses obtained with the *Assay Solution* and the *Standard Solution*, respectively.

Delete the *Test* entitled *Loss on Drying*.

Insert the following under *Tests*:

Water Determine by the *Karl Fischer Titrimetric Method*, page 552.

Change *Functional Use in Foods* to read:

Functional Use in Foods Humectant; texturizing agent; sequestrant; nutritive sweetener.

Sorbitol Solution, page 308

Delete the last sentence from the *Description*.

Insert the following under *Requirements*:

Refractive Index Between 1.455 and 1.465 at 20°.

Change the *Requirement* entitled *Water* to read:

Water Between 28.5% and 31.5%.

Replace the *Test* entitled *Assay* with the following:

Assay
Mobile Phase, Standard Solution, Resolution Solution, and *Chromatographic System* Proceed as directed in the *Assay* under *Sorbitol*.
Assay Solution Transfer an accurately weighed portion of Sorbitol Solution, equivalent to about 240 mg of sorbitol, to a 50-mL volumetric flask, dilute with water to volume, and mix.
Procedure Proceed as directed for *Procedure* in the *Assay* under *Sorbitol* and calculate the quantity, in mg, of $C_6H_{14}O_6$ in the portion of Solution taken by the formula therein given.

Insert the following under *Tests*:

Refractive Index, page 533 Determine with an Abbé or other refractometer of equal or greater accuracy.

Change *Functional Use in Foods* to read:

Functional Use in Foods Humectant; texturizing agent; sequestrant; nutritive sweetener.

Spice Oleoresins, page 310

Change the first sentence of the second paragraph of the *Description* to read:

Spice oleoresins are frequently used in commerce with added suitable food-grade diluents, preservatives, antioxidants, and other substances consistent with good manufacturing practice, as provided for under *Added Substances*, page 5.

Insert the following new monograph to precede the monograph entitled *Stearic Acid*, page 313:

Starter Distillate

Butter Starter Distillate

DESCRIPTION

Starter Distillate is the steam distillate of a culture of one or more species of *Lactococcus lactis* subsp. *diacetylactis* and/or *Leuconostoc cremoris* grown in a medium of skimmed milk that has been fortified with citric acid. It contains more than 97% water and a mixture of organic flavor compounds, principally diacetyl. It is a clear, yellowish, water-soluble liquid.

REQUIREMENTS

Assay   Not less than 90.0% and not more than 110.0% of the labeled amount of diacetyl.
Arsenic (as As)   Not more than 3 mg/kg.
Heavy Metals (as Pb)   Not more than 20 mg/kg.
Lead   Not more than 5 mg/kg.
Microbial Limits:
   Aerobic Plate Count   Not more than 10/mL.
   Coliform   Not more than 10/mL.
   Listeria sp.   Negative.
   Salmonella sp.   Negative.
   Yeasts and Molds   Not more than 10/mL.
pH   Between 2.8 and 3.8.

ADDITIONAL REQUIREMENTS

Labeling   Label to indicate the diacetyl content, in mg/mL.

TESTS

Assay   Transfer an accurately measured volume of Starter Distillate, equivalent to about 25 mg of diacetyl, to a suitable flask. Add 3 drops of phenolphthalein TS, and neutralize the acidity by titrating with 0.05 N sodium hydroxide to a faint pink endpoint. Add 0.25 mL of 30% hydrogen peroxide solution and 3 drops of 0.01% osmic acid. This is prepared by dissolving 1 g of osmium tetroxide in 1 L of water and then making a 1 in 10 dilution. (*Caution*: Osmium tetroxide and its solutions are toxic. Use proper protective equipment, and avoid contact with the eyes, skin, and clothing.) Mix, cover the flask, and allow it to stand in an incubator held at about 38° for not less than 4 h. Cool to room temperature, and titrate with 0.05 N sodium hydroxide to a faint pink endpoint. Each mL of 0.05 N sodium hydroxide is equivalent to 8.6 mg of diacetyl.
Arsenic   A 1-g sample meets the requirements of the *Arsenic Test*, page 464.
Heavy Metals   A solution of 1 g in 25 mL of water meets the requirements of the *Heavy Metals Test*, page 512, using 20 µg of lead ion (Pb) in the control (*Solution A*).
Lead   A solution of 1 g in 20 mL of water meets the requirements of the *Lead Limit Test*, page 518, using 5 µg of lead ion (Pb) in the control.
Microbial Limits:
   Aerobic Plate Count   Proceed as directed in Chapter 4 of FDA's *Bacteriological Analytical Manual*, 6th ed., Second Printing, 1989.
   Coliform   Proceed as directed in Chapter 5 of FDA's *Bacteriological Analytical Manual*, 6th ed., Second Printing, 1989.
   Listeria sp.   Proceed as directed in Chapter 29 of FDA's *Bacteriological Analytical Manual*, 6th ed., Second Printing, 1989.
   Salmonella sp.   Proceed as directed in Chapter 6 of FDA's *Bacteriological Analytical Manual*, 6th ed., Second Printing, 1989.
   Yeasts and Molds   Proceed as directed in Chapter 17 of FDA's *Bacteriological Analytical Manual*, 6th ed., Second Printing, 1989.
pH   Determine by the *Potentiometric Method*, page 531.

Functional Use in Foods   Flavoring agent.
Packaging and Storage   Store in tight containers in a cool place.

Insert the following new monograph to precede the monograph entitled *Sulfur Dioxide*, page 316:

Sucrose

Sugar; Granulated Sugar; Cane Sugar; Beet Sugar

Mol wt 342.30

DESCRIPTION

Sucrose, β-D-fructofuranosyl-α-D-glucopyranoside, is obtained for commercial use from sugar cane and sugar beets. The processed form is a white, crystalline, odorless solid with a sweet taste. It is very soluble in water, formamide, and dimethyl sulfoxide and slightly soluble in ethanol.

REQUIREMENTS

Identification

Meets the requirements under *Specific Rotation*.

Assay   Not less than 99.8 and not more than 100.2 International Sugar Degrees (°Z).
Arsenic (as As)   Not more than 1 mg/kg.

Color Not more than 75 IU.
Heavy Metals (as Pb) Not more than 5 mg/kg.
Invert Sugar Not more than 0.1%.
Lead Not more than 0.5 mg/kg.
Loss on Drying Not more than 0.1%.
Residue on Ignition Not more than 0.15%.
Specific Rotation $[\alpha]_D^{20°}$: Between +65.9° and +66.7°.

TESTS

(NOTE: Consult ICUMSA[1] rules for further details. This applies to *Assay, Color,* and *Invert Sugar.*)

Assay

Apparatus Use a saccharimeter calibrated with a certified quartz plate according to the directions of the instrument manufacturer and a 20-cm polarimeter tube with cover glasses. The tube and glasses should conform to ICUMSA specifications. Use 100-mL flasks accurate to within 0.01 mL. Maintain a water bath at 20° ± 0.1°.

Procedure Quantitatively transfer 26.000 g ± 0.002 g of the sample to the flask and add about 80 mL of water. Without heating, dissolve the sample by agitation and add water to the flask to just below the calibration mark. Place the flask in the water bath to adjust the solution to 20° ± 0.1° (degrees Centigrade unless otherwise specified). Dry the inside wall of the flask neck above the calibration mark with filter paper and, using either a hypodermic syringe or a pipet with a drawn out point, adjust to the exact volume with water. Seal the flask with a clean, dry stopper, and mix the contents thoroughly by shaking. Carefully rinse the polarimeter tube twice with two-thirds of its volume of sugar solution and fill it with sugar solution at 20° ± 0.1° in such a way that no air bubbles are trapped. Place the tube in the saccharimeter and polarize it at 20°. Determine five values to 0.05°Z, and average these values.

Arsenic A *Sample Solution* prepared using a 1-g sample, accurately weighed, meets the requirements of the *Arsenic Test,* page 464, using 1 mL of the *Standard Arsenic Solution* in the control (1 µg As).

Color

Apparatus Use a suitable variable wavelength spectrophotometer capable of measuring percent transmittance at 420 nm or a photometer with a 420- ± 10-nm band width filter. The instrument should be designed to permit the use of a 10-cm cell. When an instrument with a reference cell is used, the two cells should be identical with distilled water within ±0.2% when the instrument is set at 100% transmittance on one of the cells.

Procedure Prepare a 50% (w/w) sample solution in water. Adjust the pH to 7.0 ± 0.2 with 1% sodium hydroxide or 1% hydrochloric acid. Filter through a 0.45-µm pore-size membrane filter, using a vacuum and a diatomaceous earth filter aid (1% on solids), if necessary. Discard the first portion of the filtrate if it is cloudy. Determine the density and concentration of solids (g/mL) refractometrically. Rinse the measuring cell three times with the sample solution and then fill the cell. Measure the absorbency (A_s) at 420 nm. Calculate the color in ICUMSA units (*IU*) as follows:

$$IU = (A_s/bc) \times 1000,$$

in which *b* is the cell length, in cm, and *c* is the concentration of total solids, in g/mL, determined refractometrically and calculated from density.

Heavy Metals A solution of 4 g in 25 mL of water meets the requirements of the *Heavy Metals Test,* page 512, using 20 µg of lead ion (Pb) in the control (*Solution A*).

Invert Sugar

Apparatus Use a water bath, with vigorously boiling water, to ensure that the immersion of flasks does not interrupt the boiling. Place the flasks in the water bath so that the water level is 2 cm above the liquid surface in the flasks.

Muller's Solution Dissolve 35 g of cupric sulfate pentahydrate in 400 mL of boiling water. In a separate beaker, dissolve 173 g of potassium sodium tartrate tetrahydrate and 68 g of anhydrous sodium carbonate in 400 mL of boiling water. Cool both solutions and, while stirring, pour the sodium carbonate–potassium sodium tartrate solution into the cupric sulfate solution. Transfer the combined solutions to a 1000-mL volumetric flask, dilute to volume, and mix. Add 2 g of activated carbon, shake vigorously, and filter through hardened filter paper under vacuum. If cuprous oxide precipitates on storage, refilter the solution.

Standardized Iodine Solution Dissolve about 4.7 g of iodine in a solution of 6 g of iodate-free potassium iodide in 100 mL of water, add 3 drops of hydrochloric acid, and dilute with water to 1000 mL. Standardize to 0.0333 *N* as directed in *Iodine, 0.1* N, page 565. Adjust normality repeatedly, if necessary.

Standardized Sodium Thiosulfate Solution Dissolve about 8.7 g of sodium thiosulfate ($Na_2S_2O_3 \cdot 5H_2O$) and 67 mg of sodium carbonate in 1000 mL of freshly boiled and cooled water. Add 3 mL of 1.0 *N* sodium hydroxide. This solution contains 5.54 g of $Na_2S_2O_3$. Standardize to 0.0333 *N* as directed in *Sodium Thiosulfate, 0.1* N, page 567. Adjust normality repeatedly, if necessary.

Starch Indicator Solution Dissolve 1 g of soluble starch in 100 mL of saturated sodium chloride solution.

Procedure Transfer about 25 g of the sample, accurately weighed, into a 250-mL Erlenmeyer flask and add 100 mL of water. Dissolve and add 10 mL of *Muller's Solution* and mix well. Place the flask in a boiling water

---

[1] International Commission of Uniform Methods of Sugar Analysis (ICUMSA), 23 Avenue d'Iena, Paris 16 eme, France.

bath for 10 min ± 5 s. Remove the flask, place a small beaker over its neck, and cool rapidly, without agitation, under cold running water. Acidify the solution with 5 mL of 5 N acetic acid and immediately add an excess of *Standardized Iodine Solution* (about 20 to 40 mL). Make both of these additions without agitation to avoid the oxidation of cuprous oxide by air. Mix well and, when the precipitate is completely dissolved, titrate the excess iodine with *Standardized Sodium Thiosulfate Solution*, adding a few drops of *Starch Indicator Solution* as the endpoint is approached.

Determine a *Water Blank* by the same procedure, eliminating the sample, and a *Cold Blank* by the same procedure, but allowing the sample solution flask to stand at room temperature for 10 min rather than placing it in the boiling water bath. Calculate the percent invert sugar by the formula:

$$\% = [(V_i - V_s - B_w - B_s) \times 100]/0.2W,$$

in which V_i is the volume, in mL, of the *Standardized Iodine Solution*; V_s is the volume, in mL, of the *Standardized Sodium Thiosulfate Solution*; B_w is the volume, in mL, of the *Standardized Iodine Solution* in the *Water Blank*; B_s is the volume, in mL, of the *Standardized Iodine Solution* in the *Cold Blank*; 0.2 is a volume correction factor, in mL, used to correct for the reducing value of Sucrose; and W is the sample weight, in g.

Lead Determine as directed under *Method I* in the *Atomic Absorption Spectrophotometric Graphite Furnace Method* under the *Lead Limit Test*, page 168 of THIS SUPPLEMENT, using a 5-g sample.

Loss on Drying Dry about 5 g of the sample in a forced-draft air oven at 105° for 3 h, based on the general method, page 518.

Residue on Ignition Ignite 1 g as directed in the general method, page 533.

Specific Rotation, page 530 Dissolve 26 g of the sample in water and dilute to 100 mL at 20°. Determine the specific rotation using a 20-cm polarimeter tube.

Functional Use in Foods Nutritive sweetener; formulation and texturizing aid.

Packaging and Storage Store in tight containers in a dry place.

Sulfuric Acid, page 317

Change the *Test* entitled *Selenium* to read:

Selenium Determine as directed in *Method II* under the *Selenium Limit Test*, page 537, using 300 mg of sample.

The absorbance of the extract from the *Sample Preparation* is not greater than that from the *Standard Preparation*.

L-Threonine, page 326

Change the second sentence in the *Description* to read:

It is freely soluble in water, but insoluble in alcohol, in ether, and in chloroform.

Change the *Requirements* for *Assay*, *Arsenic*, and *Specific Rotation* to read:

Assay Not less than 98.5% and not more than 101.5% $C_4H_9NO_3$, calculated on the dried basis.
Arsenic (as As) Not more than 1.5 mg/kg.
Specific Rotation $[\alpha]_D^{20°}$: Between −26.5° and −29.0°, calculated on the dried basis.

Change the *Test* entitled *Arsenic* to read:

Arsenic A *Sample Solution* prepared as directed for organic compounds meets the requirements of the *Arsenic Test*, page 464, using 1.5 mL of the *Standard Arsenic Solution* in the control (1.5 µg As).

DL-Tryptophan, page 339

Change the *Requirements* entitled *Assay*, *Arsenic*, and *Heavy Metals* to read:

Assay Not less than 98.5% and not more than 101.5% $C_{11}H_{12}N_2O_2$, calculated on the dried basis.
Arsenic (as As) Not more than 1.5 mg/kg.
Heavy Metals (as Pb) Not more than 20 mg/kg.

Delete the *Requirements* entitled *Ammonium Salts*, *Chloride*, *Indole*, *Iron*, *Nitrogen*, and *Sulfate*.

Delete the *Tests* entitled *Ammonium Salts*, *Chloride*, *Indole*, *Iron*, *Nitrogen*, and *Sulfate*.

Change the *Tests* entitled *Arsenic* and *Heavy Metals* to read:

Arsenic A *Sample Solution* prepared as directed for organic compounds meets the requirements of the *Arsenic Test*, page 464, using 1.5 mL of the *Standard Arsenic Solution* in the control (1.5 µg As).
Heavy Metals Prepare and test a 1-g sample as directed under the *Heavy Metals Limit Test*, page 512, using 20 µg of lead ion (Pb) in the control (*Solution A*).

L-Tryptophan, page 340

Change the *Requirements* entitled *Assay* and *Arsenic* to read:

Assay   Not less than 98.5% and not more than 101.5% $C_{11}H_{12}N_2O_2$, calculated on the dried basis.
Arsenic (as As)   Not more than 1.5 mg/kg.

Delete the *Requirement* entitled *Ammonium Salts*.

Delete the *Test* entitled *Ammonium Salts*.

Change the *Test* entitled *Arsenic* to read:

Arsenic   A *Sample Solution* prepared as directed for organic compounds meets the requirements of the *Arsenic Test*, page 464, using 1.5 mL of the *Standard Arsenic Solution* in the control (1.5 µg As).

L-Tyrosine, page 341

Change the *Requirements* entitled *Assay*, *Arsenic*, and *Specific Rotation* to read:

Assay   Not less than 98.5% and not more than 101.5% $C_9H_{11}NO_3$, calculated on the dried basis.
Arsenic (as As)   Not more than 1.5 mg/kg.
Specific Rotation $[\alpha]_D^{20°}$: Between −11.3° and −12.3°, calculated on the dried basis.

Delete the *Requirements* for *Iron* and *Nitrogen*.

Change the *Tests* entitled *Arsenic* and *Specific Rotation* to read:

Arsenic   A *Sample Solution* prepared as directed for organic compounds meets the requirements of the *Arsenic Test*, page 464, using 1.5 mL of the *Standard Arsenic Solution* in the control (1.5 µg As).
Specific Rotation   Determine in a solution containing 5 g of a previously dried sample in sufficient 1 *N* hydrochloric acid to make 100 mL.

Delete the *Tests* for *Iron* and *Nitrogen*.

Insert the following new monograph to precede the monograph entitled L-*Valine*, page 341:

Urea

Carbamide

CH_4N_2O Mol wt 60.06

DESCRIPTION

A colorless to white, prismatic, crystalline powder or small, white pellets. It is practically odorless, but upon standing may develop a slight odor of ammonia. It is freely soluble in water and in boiling alcohol and practically insoluble in chloroform and in ether. It is commonly produced from CO_2 by ammonolysis or from cyanamide by hydrolysis. It melts at a range of 132° to 135°.

REQUIREMENTS

Identification

A. Heat about 500 mg in a test tube until it liquifies. Ammonia vapor is produced. Continue heating until the liquid becomes turbid, and then cool. Dissolve the fused mass in a mixture of 10 mL of water and 1 mL of sodium hydroxide solution (1 in 10). Add 1 drop of cupric sulfate TS. A reddish violet-colored solution is produced.
B. Dissolve 100 mg in 1 mL of water, and add 1 mL of nitric acid. A white precipitate of urea nitrate is produced.

Assay   Not less than 99.0% and not more than 100.5% CH_4N_2O.
Alcohol-Insoluble Matter   Not more than 0.04%.
Arsenic (as As)   Not more than 3 mg/kg.
Chloride   Not more than 0.007%.
Heavy Metals (as Pb)   Not more than 10 mg/kg.
Lead   Not more than 5 mg/kg.
Loss on Drying   Not more than 1.0%.
Residue on Ignition   Not more than 0.1%.
Sulfate   Not more than 0.01%.

TESTS

Assay   Transfer about 500 mg, accurately weighed, to a 200-mL volumetric flask, and dissolve in 100 mL of water, dilute to volume with water, and mix. Pipet 2 mL of this solution into a semimicro Kjeldahl digestion flask, and proceed as directed under *Nitrogen Determination (Method II)*, page 521. Heat the sample until it begins to fume, and then heat for 1 additional h. Each mL of 0.01 *N* acid is equivalent to 0.3003 mg of CH_4N_2O.
Alcohol-Insoluble Matter   Dissolve about 5 g in 50 mL of warm alcohol. If any residue remains, filter the solution

through a tared filter, wash the residue, and filter with 20 mL of warm alcohol. Dry at 105° for 1 h. Cool in a desiccator, and weigh.

Arsenic A solution of 1 g in 10 mL of water meets the requirements of the *Arsenic Test*, page 464.

Chloride A 0.2-g sample meets the requirements of the *Chloride Limit Test*, page 471, using 14 µg of chloride ion (Cl) in the control.

Heavy Metals Prepare and test a 2-g sample as directed under the *Heavy Metals Test*, page 512, using 20 µg of lead ion (Pb) in the control (*Solution A*).

Lead Prepare and test a 2-g sample as directed under the *Lead Limit Test*, page 518, using 10 µg of lead ion (Pb) in the control.

Loss on Drying, page 518 Dry at 105° for 3 h.

Residue on Ignition Ignite 1 g as directed under *Method I* in the general method, page 533.

Sulfate A 2-g sample meets the requirements of the *Sulfate Limit Test*, page 471, using 200 µg of sulfate ion (SO_4) in the control.

Functional Use in Foods Formulation aid; fermentation aid.

Packaging and Storage Store in a well-closed container.

L-Valine, page 341

Change the *Requirements* entitled *Assay*, *Arsenic*, and *Specific Rotation* to read:

Assay Not less than 98.5% and not more than 101.5% $C_5H_{11}NO_2$, calculated on the dried basis.

Arsenic (as As) Not more than 1.5 mg/kg.

Specific Rotation $[\alpha]_D^{20°}$: Between +27.2° and +29.0°, calculated on the dried basis.

Change the *Test* entitled *Arsenic* to read:

Arsenic A *Sample Solution* prepared as directed for organic compounds meets the requirements of the *Arsenic Test*, page 464, using 1.5 mL of the *Standard Arsenic Solution* in the control (1.5 µg As).

Xanthan Gum, page 347

Change the *Requirements* entitled *Assay*, *Ash*, and *Loss on Drying* to read:

Assay It yields, on the dry basis, not less than 4.2% and not more than 5.4% carbon dioxide (CO_2), corresponding to between 91.0% and 117.0% of Xanthan Gum.

Ash Between 6.5% and 16.0%.

Loss on Drying Not more than 15.0%.

Change the second sentence of the *Test* entitled *Viscosity* to read:

Determine the viscosity of one solution at 23.9° (75°F) as directed under *Viscosity of Sodium Carboxymethylcellulose*, page 550, using a No. 3 spindle rotating at 60 rpm (Brookfield or equivalent).

Xylitol, page 348

Insert the following as the second sentence of the *Description*:

Xylitol is found in most fruits and berries, as well as in vegetables.

Change the first sentence of the *Identification* to read:

The infrared absorption spectrum of a potassium bromide dispersion of Xylitol exhibits maxima only at the same wavelengths as those of a similar preparation of USP Xylitol Reference Standard.

Change the *Requirement* entitled *Assay* to read:

Assay Not less than 98.5% and not more than 101.0% $C_5H_{12}O_5$, calculated on the anhydrous basis.

Delete the *Requirement* entitled *Loss on Drying*.

Insert the following under *Requirements*:

Water Not more than 0.5%.

Replace the *Test* entitled *Assay* as follows:

Assay

Internal Standard Solution Transfer about 500 mg of erythritol, accurately weighed, into a 25-mL volumetric flask, dilute to volume with water, and mix.

Standard Solution Transfer about 25 mg each of L-arabinitol, galactitol, mannitol, and sorbitol, accurately weighed, to a 100-mL volumetric flask, dilute to volume with water, and mix. To an accurately measured volume of this solution, add an accurately weighed amount of USP Xylitol Reference Standard to obtain a solution with a known concentration of about 49 mg/mL.

Assay Preparation Transfer about 5 g of the sample, accurately weighed, into a 100-mL volumetric flask, dilute to volume with water, and mix.

Chromatographic System Use a gas chromatograph equipped with a flame-ionization detector and a 2-m × 2-mm glass column packed with 3% liquid phase of 25% phenyl–25% cyanopropylmethylsilicone (OV-225 or equivalent) on silanized siliceous earth support (Chromosorb W-HP or equivalent). The carrier gas is nitrogen flowing at about 30 mL/min. The injector port temperature is 250°, the column temperature is 200°, and the detector temperature is 250°. Chromatograph the derivatized *Standard Solution* prepared as directed under *Procedure*, and record the peak responses. The relative retention times corresponding to erythritol, L-arabinitol, xylitol, galactitol, mannitol, and sorbitol are usually about 1.0, 2.77, 3.90, 6.96, 7.63, and 8.43, respectively. The relative standard deviation of the response ratios of the derivatized xylitol to the derivatized erythritol from three replicate injections does not exceed 2.0%.

Procedure Pipet 1-mL portions of the *Standard Solution* and the *Assay Preparation* into separate 100-mL round-bottom boiling flasks. To each flask, add 1.0 mL of *Internal Standard Solution*, and evaporate the respective mixtures to dryness on a water bath at 60° with the aid of a rotary evaporator. Dissolve each dry residue in 1 mL of pyridine, and add 1 mL of acetic anhydride to each flask. Boil each solution under reflux for 1 h to complete the acetylation. Separately inject 1-µL portions of the derivatized solutions from the *Assay Preparation* and the *Standard Solution* into the gas chromatograph and measure the peak responses. Calculate the percentage of Xylitol, on the as-is basis, by the formula:

$$100(W_S/W_U)(R_U/R_S),$$

in which W_S is the weight, in mg, of USP Xylitol Reference Standard used for the *Standard Solution*; W_U is the weight, in mg, of the sample taken for the *Assay Preparation*; and R_U and R_S are the ratios of peak responses of the derivatized analyte to the derivatized erythritol from the *Internal Standard Solution* obtained from the *Assay Preparation* and the *Standard Solution*, respectively. Using the value obtained in the *Water* determination, correct the percentage to the anhydrous basis.

Delete the *Test* entitled *Loss on Drying*.

Replace the *Test* entitled *Other Polyols* with the following:

Other Polyols

Internal Standard Solution, Standard Solution, Assay Preparation, and Chromatographic System Proceed as directed in the *Assay*.

Procedure Proceed as described in the *Assay*. Calculate the percentage of each polyol—L-arabinitol, galactitol, mannitol, and sorbitol—by the formula therein given, in which W_S refers to the weight, in mg, of the respective polyol taken for the *Standard Solution*; R_S is the peak response ratio of the corresponding polyol obtained from the *Standard Solution;* and R_U is the peak response ratio of the corresponding polyol obtained from the *Assay Preparation*. Sum the four individual polyol percentages to obtain the total.

Insert the following under *Tests*:

Water Determine water content as directed under the *Karl Fischer Titrimetric Method*, page 552.

Change *Packaging and Storage* to read:

Packaging and Storage Store in well-closed containers in a dry place.

Insert the following new monograph to precede the monograph entitled *Zinc Gluconate*, page 349:

Zein

DESCRIPTION

Zein is a very light yellow- to straw-colored, water-insoluble, granular or fine powder comprising the prolamine protein component of corn (*Zea mays* Linne'). It is produced commercially by extraction from corn gluten with alkaline aqueous isopropyl alcohol. The extract is then cooled, which causes the Zein to precipitate.

REQUIREMENTS

Identification

A. Dissolve about 0.1 g in 10 mL of 0.1 N sodium hydroxide, and add a few drops of cupric sulfate TS. Warm in a water bath: a purple color develops.
B. To a test tube containing 25 mg of the sample, add 1 mL of nitric acid. Agitate vigorously: the solution becomes light yellow. Further addition of about 10 mL of 6 N ammonium hydroxide produces an orange color.

Assay Not less than 88.0% and not more than 96.0% protein, calculated on a dried basis.
Arsenic (as As) Not more than 3 mg/kg.
Heavy Metals (as Pb) Not more than 20 mg/kg.
Lead Not more than 5 mg/kg.
Loss on Drying Not more than 8.0%.
Loss on Ignition Not more than 2%.

TESTS

Assay   Proceed as directed under *Nitrogen Determination*, page 521. Calculate the percent protein (*P*) by the formula:

$$P = 6.25N,$$

in which *N* is the percent nitrogen.

Arsenic   A *Sample Solution* prepared as directed for organic compounds meets the requirements of the *Arsenic Test*, page 464.

Heavy Metals   Prepare and test a 1-g sample as directed under *Method II* under the *Heavy Metals Test*, page 513, using 20 µg of lead ion (Pb) in the control (*Solution A*).

Lead   A *Sample Solution* prepared as directed for organic compounds meets the requirements of the *Lead Limit Test*, page 518, using 5 µg of lead ion (Pb) in the control.

Loss on Drying, page 518   Dry a 2-g sample in an air oven at 105° for 2 h.

Loss on Ignition   Proceed as directed under *Ash (Total)*, page 466, using a 2-g sample.

Functional Use in Foods   Surface-finishing agent.
Packaging and Storage   Store in well-closed containers.

Zinc Gluconate, page 349

Change the *Description* to read:

Zinc gluconate is a white or nearly white, granular or crystalline powder. It is freely soluble in water and very slightly soluble in alcohol.

Change the *Requirement* entitled *Water* to read:

Water   *Powder or granular*: not less than 5.0% and not more than 10.0%; *trihydrate*: not less than 8.0% and not more than 11.6%.

Insert the following heading after the *Requirements* section:

ADDITIONAL REQUIREMENTS

Replace the *Additional Requirement* entitled *Labeling* with the following:

Labeling   Label to indicate the powder or granular form of the product.

Change the first sentence of the *Test* entitled *Lead, Procedure*, to read:

Using a Perkin-Elmer Model 403 (or equivalent) atomic absorption spectrophotometer equipped with a deuterium arc background corrector, a digital readout device, and a burner head capable of handling 10% solids, blank the instrument with water following the manufacturer's operating instructions.

Change the *Test* entitled *Water* to read:

Water   Determine by the *Karl Fischer Titrimetric Method*, page 552, using the *Residual Titration* procedure.

3/ Specifications for Flavor Aromatic Chemicals and Isolates

Benzyl Cinnamate, page 358
[FEMA No. 2142]

Change the *Solidification Point* from 33° and 34.5° to 33° and 35.0°.

Benzyl Isobutyrate, page 358
[FEMA No. 2141]

Change the *Ref. Index* from 1.489–1.492 to 1.488–1.492.

Change the *Sp. Gr.* from 1.001–1.005 to 1.000–1.005.

Benzyl Isovalerate, page 358
[FEMA No. 2152]

Change the *Sp. Gr.* from 0.985–0.991 to 0.983–0.989.

2,6-Dimethyl-5-heptenal, page 66, Second Supplement
[FEMA No. 2389]

Change the *Assay* from 85.0% min (M-8a) to 85.0% min (M-3a as $C_9H_{16}O$; 1.0 g/14.01).

Change the *A.V. Max* from 1.0 to 5.0.

Change the *Ref. Index* from 1.443–1.448 to 1.442–1.447.

Change the *Sp. Gr.* from 0.852–0.858 to 0.848–0.854.

Estragole, page 372
[FEMA No. 2411]

Under *Assay*, insert 95% min (M-8a).

Ethyl Formate, page 376
[FEMA No. 2434]

Change the *Acidity* from 0.1% to 0.2%.

Isoamyl Butyrate, page 388
[FEMA No. 2060]

Change the *Sp. Gr.* from 0.860–0.865 to 0.861–0.866.

Isoamyl Formate, page 388
[FEMA No. 2069]

Change the *A.V. Max* from 1.0 to 3.0.

Change the *Sp. Gr.* from 0.878–0.885 to 0.881–0.889.

Isovaleric Acid, page 392
[FEMA No. 3102]

Change the *Ref. Index* from 1.403–1.405 to 1.401–1.405.

Change the *Sp. Gr.* from 0.928–0.931 to 0.923–0.928.

α-Pinene, page 412
[FEMA No. 2902]

Include as a *Synonym* L-α-Pinene.

Change the *Sp. Gr.* from 0.851–0.855 to 0.855–0.860.

Under *Other Requirements*, page 413, change *Angular Rotation* from NLT –40° to: –35° to –50°.

β-Pinene, page 412
[FEMA No. 2903]

Change the *Sp. Gr.* from 0.864–0.868 to 0.867–0.871.
Under *Other Requirements*, page 413, change *Angular Rotation* from NLT –20° to: –15° to –30°.

γ-Undecalactone, page 418
("Aldehyde C-14")
[FEMA No. 3091]

Change the *Ref. Index* from 1.450–1.454 to 1.448–1.453.

Change the *Sp. Gr.* from 0.942–0.945 to 0.940–0.945.

New Flavor Monographs

General Information and Description

Name of Substance (Synonyms)	Mol Wt/Formula/ Structure	Physical Form/Odor[1]	Solubility/ B.P.	GLC Profile	Solubility in Alcohol
Anisyl Formate (*p*-Methoxybenzyl Formate) [FEMA No. 2110]	166.18/$C_9H_{10}O_3$	Colorless to pale yel liq/ sweet, floral, tonka-like			
Benzyl Formate [FEMA No. 2145]	136.15/$C_8H_8O_2$	Colorless to pale yel liq/ sweet, balsamic, floral			
Butyl Stearate (Butyl Octadecanoate) [FEMA No. 2214]	340.57/$C_{22}H_{44}O_2$	Colorless, waxy solid/ odorless to faintly fatty			
D-**Camphor** [FEMA No. 2230]	152.23/$C_{10}H_{16}O$	White to gray translucent cryst or fused mass/characteristic			
L-**Carveol** (*p*-Mentha-6,8-dien-2-ol) [FEMA No. 2247]	152.23/$C_{10}H_{16}O$	Colorless to pale yel liq/ spearminty			
L-**Carvyl Acetate** (*p*-Mentha-6,8-dien-2-yl Acetate) [FEMA No. 2250]	194.27/$C_{12}H_{18}O_2$	Colorless to pale yel liq/ spearminty			
Cinnamyl Butyrate [FEMA No. 2296]	204.27/$C_{13}H_{16}O_2$	Colorless to pale yel liq/ fruity, balsamic			
Cinnamyl Cinnamate [FEMA No. 2298]	264.31/$C_{18}H_{16}O_2$	Mixture of *cis* and *trans* isomers, low-melting solid			
Cinnamyl Isobutyrate [FEMA No. 2297]	204.27/$C_{13}H_{16}O_2$	Colorless to pale yel liq/ sweet, balsamic, fruity			
γ-**Decalactone** (4-Hydroxydecanoic Acid Lactone) [FEMA No. 2361]	170.25/$C_{10}H_{18}O_2$	Colorless to pale yel liq/ fruity, peach-like			
Dimethyl Sulfide (Methyl Sulfide Thiobismethane) [FEMA No. 2746]	62.13/C_2H_6S	Colorless to pale yel liq/ disagreeable, intense boiled cabbage			
γ-**Dodecalactone** (4-Hydroxydodecanoic Acid Lactone) [FEMA No. 2400]	198.31/$C_{12}H_{22}O_2$	Colorless to pale yel liq/ fruity, peach-like, pear-like			

Requirements

I.D. Test	Assay Min, %[2]	A.V. Max	Ref. Index	Sp. Gr.	Other Requirements
	90.0% (M-8b)	3.0	1.521–1.525	1.138–1.142	
	92.0% (M-8b)	3.0	1.508–1.515	1.082–1.092	
					Melting Range: 17° to 21° (p. 519); **Iodine Value:** 1 max; **Saponification Value:** 165 to 180
	99.0% (M-8b)				**Melting Range:** 174° to 179° (p. 519); **Angular Rotation:** +41° to +43°
	95.0% (M-8b) cis isomer 45% ± 5% trans isomer 55% ± 5%		1.493–1.497	0.947–0.953	**Angular Rotation:** −117° to −130°
	98.0% (M-6; 1.2 g/97.14)	1.0	1.473–1.479	0.964–0.970	**Angular Rotation:** 1.2 −90° to −120°
	96.0% (M-6; 1.4 g/102.14)	1.0	1.525–1.530	1.010–1.015	
	95.0% (M-6; 1.6 g/123.16)	2.0			
	96.0% $C_{13}H_{16}O_2$ (M-6; 1.4 g/ 102.1)	3.0	1.523–1.528	1.006–1.009	
	97.0% (M-8a)	1.0	1.447–1.451	0.950–0.955	
	98.0% (M-8a)		1.431–1.441	0.842–0.847	
	97.0% (M-8a)	1.0	1.451–1.456	0.933–0.938	

New Flavor Monographs *Continued*

General Information and Description

Name of Substance (Synonyms)	Mol Wt/Formula/ Structure	Physical Form/Odor[1]	Solubility/ B.P.	GLC Profile	Solubility in Alcohol
Ethylene Brassylate [FEMA No. 3543]	270.37/$C_{15}H_{26}O_4$	Colorless to pale yel liq/ sweet, musky			
4-Ethylguaiacol (4-Hydroxy-3-methoxy-ethylbenzene) [FEMA No. 2436]	152.19/$C_9H_{12}O_2$	Colorless to pale yel liq/ warm, spicy, medicinal			
Ethyl-3-Methyl-thiopropionate [FEMA No. 3343]	148.23/$C_6H_{12}O_2S$				
Ethyl Oleate (Ethyl 9-Octadecenoate) [FEMA No. 2450]	310.52/$C_{20}H_{38}O_2$	Colorless to pale yel liq/ floral			
Glyceryl Tri-propanoate (Tripropionin) [FEMA No. 3286]	260.29/$C_{12}H_{20}O_6$	Colorless to pale yel liq/ odorless with a bitter taste			
γ-Hexalactone (4-Hydroxyhexanoic Acid Lactone) [FEMA No. 2556]	114.15/$C_6H_{10}O_2$	Colorless to pale yel liq/ herbaceous, sweet			
cis-3-Hexenyl Acetate [FEMA No. 3171]	142.19/$C_8H_{14}O_2$	Colorless to pale yel liq/ powerful green note			
trans-2-Hexenyl Acetate [FEMA No. 2564]	142.19/$C_8H_{14}O_2$	Colorless to pale yel liq/ green note			
4-Hydroxy-2,5-dimethyl-3(2H)furanone [FEMA No. 3174]	128.13/$C_6H_8O_3$	White to pale yel solid/ fruity, caramel, burnt sugar			
Isoborneol [FEMA No. 2158]	154.24/$C_{10}H_{18}O$	White cryst solid/piney, camphoraceous			
2-Methylpentanoic Acid [FEMA No. 2754]	116.16/$C_6H_{12}O_2$	Colorless to pale yel liq/ caramel, pungent			
2-Methyl-2-pentenoic Acid [FEMA No. 3195]	114.04/$C_6H_{10}O_2$	Colorless to pale yel liq			
4-Methylpentanoic Acid [FEMA No. 3463]	116.16/$C_6H_{12}O_2$	Colorless to pale yel liq/ sour, penetrating			

FCC III-THIRD SUPPLEMENT / *Specifications for Flavor* / 161

Requirements

I.D. Test	Assay Min, %[2]	A.V. Max	Ref. Index	Sp. Gr.	Other Requirements
	97.0% $C_{15}H_{26}O_4$ (M-6; 1.1 g/67.59)	1.0	1.468–1.473	1.040–1.045	
	98.0% (M-8b)		1.525–1.530	1.061–1.064	
			1.457–1.463	1.030–1.035	
		1.0	1.448–1.453	0.868–0.873	**Saponification Value**: 175 to 190
	97.0% $C_{12}H_{20}O_6$ (M-6; 1.5 g/43.38)	2.0	1.431–1.435	1.078–1.082	
	98.0% (M-8a)	1.0	1.437–1.442	1.020–1.025	
	98.0% as sum of isomers with major peak (*cis*) at 92% min (M-8a)	1.0	1.425–1.429	0.896–0.901	
	98.0% as sum of *cis* and *trans* isomers not less than 90% (M-8a)		1.425–1.430	0.890–0.897	
	98.0% (M-8a in a suitable solvent)				
	92.0% (M-8a in a suitable solvent)				**Melting Range**: NLT 206° (p. 519)
	98.0% (M-11a; 2.0 g/58.08)		1.411–1.416	0.919–0.922	
	98.0% (M-11a; 2.0 g/57.02)		1.450–1.460	0.976–0.982	
	98.0% (M-11a; 2.0 g/58.08)		1.412–1.417	0.919–0.926	

New Flavor Monographs Continued

General Information and Description

Name of Substance (Synonyms)	Mol Wt/Formula/ Structure	Physical Form/Odor[1]	Solubility/ B.P.	GLC Profile	Solubility in Alcohol
3-Methylthiopropion-aldehyde (Methional) [FEMA No. 2747]	104.17/C_4H_8OS	Colorless to pale yel liq/ meaty, potato			
Myristaldehyde (Tetradecanal) [FEMA No. 2763]	212.36/$C_{14}H_{28}O$	Colorless to pale yel liq/ fatty, orris-like			
Neryl Acetate (cis-3,7-Dimethyl-2,6-octadien-1-yl-Acetate) [FEMA No. 2773]	196.29/$C_{12}H_{20}O_2$	Colorless to pale yel liq/ sweet, floral			
2-Nonanone (Methyl Heptyl Ketone) [FEMA No. 2785]	142.24/$C_9H_{18}O$	Colorless to pale yel liq/ fruity, floral, fatty, herbaceous			
1-Octene-3-ol (Amyl Vinyl Carbinol) [FEMA No. 2805]	128.21/$C_8H_{16}O$	Colorless to pale yel liq/mushroom-like, herbaceous			
Octyl Isobutyrate (Octyl 2-Methylpropanoate) [FEMA No. 2808]	200.33/$C_{12}H_{24}O_2$	Colorless to pale yel liq/ refreshing, herbaceous			
Phenylethyl Anthranilate [FEMA No. 2859]	241.29/$C_{15}H_{15}NO_2$	Colorless to pale yel cryst mass/neroli-like, grape undertone			
Phenylethyl Butyrate [FEMA No. 2861]	192.26/$C_{12}H_{16}O_2$	Colorless to pale yel liq/ green, hay-like			
Piperidine (Hexahydropyridine) [FEMA No. 2908]	85.15/$C_5H_{11}N$	Colorless to pale yel liq/ ammoniacal, fishy, nauseating			
Tetrahydrofurfuryl Alcohol [FEMA No. 3056]	102.13/$C_5H_{10}O_2$	Colorless liq/mild, warm, oily, caramel			
Thymol [FEMA No. 3066]	150.22/$C_{10}H_{14}O$	White cryst/phenol-like			
Tolualdehyde (mixed isomers) (Tolyl Aldehyde, mixed isomers, Methyl Benzaldehyde) [FEMA No. 3068]	120.15/C_8H_8O	Colorless liq/cherry-like			

Requirements

I.D. Test	Assay Min, %[2]	A.V. Max	Ref. Index	Sp. Gr.	Other Requirements
	98.0% (M-8a)		1.484–1.493	1.038–1.048	
	85.0% (M-4; 1.5 g/106.18)	5.0	1.438–1.445	0.825–0.830	
	96.0% $C_{12}H_{20}O_2$ (M-6; 1 g/98.15) Predominantly *cis* isomer by M-8a	1.0	1.458–1.464	0.905–0.914	
	97.0% (M-8a)		1.418–1.423	0.817–0.823	
	98.0% (M-8a)		1.434–1.442	0.831–0.839	
	98.0% (M-8a)	1.0	1.420–1.425	0.853–0.858	
	98.0% (M-8b)	1.0			**Solidification Point**: NLT 40°
	98.0% (M-8b)	1.0	1.487–1.492	0.991–0.995	
	98.0% (M-8a)		1.450–1.454	0.858–0.862	
	99.0% (M-8a)		1.452–1.453	1.050–1.052	
	99.0% (M-8b)				**Melting Range**: 48° to 51° (p. 519)
	94.0% (M-2a; 1.2 g/60.08)	5.0	1.540–1.549	1.019–1.029	

New Flavor Monographs Continued

General Information and Description

Name of Substance (Synonyms)	Mol Wt/Formula/ Structure	Physical Form/Odor[1]	Solubility/ B.P.	GLC Profile	Solubility in Alcohol
para-Tolualdehyde (*p*-Tolyl Aldehyde, *p*-Methylbenzaldehyde) [FEMA No. 3068]	120.15/C_8H_8O	Colorless liq/cherry-like			
Trimethylamine [FEMA No. 3241]	59.11/C_3H_9N	Gas/pungent, fishy, ammoniacal			
2-Undecanone (Methyl Nonyl Ketone) [FEMA No. 3093]	170.30/$C_{11}H_{22}O$	Colorless to pale yel liq/ citrus, fatty, rue-like			
Valeraldehyde [FEMA No. 3098]	86.13/$C_5H_{10}O$	Colorless to pale yel liq/ characteristic			

NOTES:

[1]cryst = crystal or crystalline; liq = liquid; NLT = not less than; yel = yellow.

[2]Sample weight and equivalence factor are given for methods M-2a, M-4, and M-6; sample weights (g) and amounts (mg) of substance equivalent to 0.5 N sodium hydroxide are given for method M-11a.

Requirements

I.D. Test	Assay Min, %[2]	A.V. Max	Ref. Index	Sp. Gr.	Other Requirements
	97.0% (M-2a; 1.2 g/60.08)	5.0	1.542–1.548	1.012–1.018	
	98.0% (M-8a in suitable solvent)				
	96.0% (M-8a)	5.0	1.428–1.432	0.822–0.826	
	97.0% (M-4; 1 g/43.07)	5.0	1.390–1.395	0.805–0.809	

4/ Test Methods for Flavor Aromatic Chemicals and Isolates

No change.

5/ GLC Analysis of Flavor Aromatic Chemicals and Isolates

No change.

6/ General Tests and Apparatus

Insert the following new test at the end of the *Lead Limit Test* and to precede the test entitled *Loss on Drying*, page 518:

Atomic Absorption Spectrophotometric Graphite Furnace Method

Lead This atomic absorption spectrophotometric method employs a graphite furnace and is primarily intended for the analysis of substances containing less than 1 mg/kg of lead. *Method I* is intended for substances that are soluble in water, such as sugars, whereas *Method II* is for those substances immiscible with water, such as edible oils. Unless stated otherwise in the monograph, use *Method I*.

Apparatus Use a suitable atomic absorption spectrophotometer (Perkin-Elmer Model 3100 or equivalent) fitted with a graphite furnace (Perkin-Elmer HGA 600 or equivalent). Use a lead hollow cathode lamp (Perkin-Elmer or equivalent) with argon as the carrier gas. Follow the manufacturers' directions for setting the appropriate instrument parameters for lead determination.

(NOTE: For this test, use reagent-grade chemicals with as low a lead content as is practicable, as well as high-purity water and gases. Before use in this analysis, rinse all glassware and plasticware twice with 10% nitric acid and twice with 10% hydrochloric acid, and then rinse them thoroughly with high-purity water, preferably obtained from a mixed-bed strong-acid, strong-base ion-exchange cartridge capable of producing water with an electrical resistivity of 12 to 15 megohms.)

Method I

Hydrogen Peroxide–Nitric Acid Solution Dissolve equal volumes of 10% hydrogen peroxide and 10% nitric acid. (NOTE: Use caution.)

Lead Nitrate Stock Solution Dissolve 159.8 mg of ACS reagent-grade lead nitrate (alternatively, use NIST Standard Reference Material, containing 10 mg of lead per kg, or equivalent) in 100 mL of *Hydrogen Peroxide–Nitric Acid Solution*. Dilute to 1000.0 mL with *Hydrogen Peroxide–Nitric Acid Solution*, and mix. Prepare and store this solution in glass containers that are free from lead salts. Each mL of this solution contains the equivalent of 100 µg of lead ion.

Standard Lead Solution On the day of use, dilute 10.0 mL of *Lead Nitrate Stock Solution* with *Hydrogen Peroxide–Nitric Acid Solution* to 100.0 mL, and mix. Each mL of *Standard Lead Solution* contains the equivalent of 10 µg of lead ion.

Standard Solutions Prepare a series of lead standard solutions serially diluted from the *Standard Lead Solution* in *Hydrogen Peroxide–Nitric Acid Solution*. Into separate 100-mL volumetric flasks containing 1 mL of nitric acid and 1 mL of 30% hydrogen peroxide, pipet 0.2, 0.5, 1, and 2 mL, respectively, of *Standard Lead Solution*, dilute to volume with *Hydrogen Peroxide–Nitric Acid Solution*, and mix. The *Standard Solutions* contain, respectively, 0.02, 0.05, 0.1, and 0.2 µg of lead per mL. (For lead limits greater than 1 mg/kg, prepare a series of standard solutions in a range encompassing the expected lead concentration in the sample.)

Sample Solution (NOTE: Perform this procedure in a fume hood, and wear safety glasses.) Accurately weigh 1 g of the sample, and place it in a large test tube. Add 1 mL of nitric acid. Place the test tube in a rack in a boiling water bath. As soon as the rusty tint is gone, add 1 mL of 30% hydrogen peroxide dropwise to avoid a vigorous reaction, and wait for bubbles to form. Stir with an acid-washed plastic spatula if necessary. Remove the test tube from the water bath, and let it cool. Transfer the solution to a 10-mL volumetric flask, and dilute to volume with *Hydrogen Peroxide–Nitric Acid Solution*, and mix. Use this solution for analysis.

Method II

Butanol–Nitric Acid Solution Slowly add 50 mL of nitric acid to approximately 500 mL of butanol in a 1000-mL volumetric flask. Dilute to volume with butanol, and mix.

Standard Solutions Prepare as directed for *Standard Solutions* under *Method I*, using *Butanol–Nitric Acid Solution* instead of *Hydrogen Peroxide–Nitric Acid Solution*. Do not add 1 mL of nitric acid or 1 mL of 30% hydrogen peroxide to this solution.

Sample Solution (NOTE: Perform this procedure in a fume hood, and wear safety glasses.) Transfer about 1 g of the sample, accurately weighed, to an accurately tared 10-mL volumetric flask. Reweigh the flask to obtain the exact weight of the sample. Samples are most easily transferred as liquids; therefore, samples solid at room temperature should be melted on a steam bath. Stir liquified samples with an acid-washed plastic spatula if necessary. Heat the sample on a steam bath to assist dissolution if necessary. Dilute to volume with *Butanol–Nitric Acid Solution*, and mix. Use this solution for analysis.

Procedure

Tungsten Solution Transfer 0.1 g of tungstic acid (H_2WO_4) and 5 g of sodium hydroxide pellets into a 50-mL plastic bottle. Add 5.0 mL of high-purity water, and mix. Heat the mixture in a hot water bath until complete solution is achieved. Cool, and store at room temperature.

Procedure Place the graphite tube in the furnace. Inject a 20-µL aliquot of the *Tungsten Solution* into the graphite tube, using the following sequence of conditions: dry at 110° for 20 s, char at 700° to 900° for 20 s, and atomize at 2700° for 10 s; repeat this procedure once more using a second 20-µL aliquot of the *Tungsten Solution*. Clean the quartz windows.

Standard Curve (NOTE: The sample injection technique is the most crucial step in controlling the precision of the analysis; the volume of the sample must remain constant. Rinse the µL pipet tip (Eppendorf or equivalent) three times with either the *Standard Solutions* or *Sample Solution* before injection. Use a fresh pipet tip for each injection, and start the atomization process immediately after injecting the sample. Between injections, flush the graphite tube of any residual lead by purging at a high temperature as recommended by the manufacturer.) With the hollow cathode lamp properly aligned for maximum absorbance and the wavelength set at 283.3 nm, atomize 20-µL aliquots of the four *Standard Solutions*, using the following sequence of conditions: dry at 110° for 30 s, with a 20-s ramp period and a 10-s hold time, then char at 700° for 42 s, with a 20-s ramp period and a 22-s hold time, and then atomize at 2300° for 7 s.

Plot a standard curve using the concentration, in µg/mL, of each *Standard Solution* versus its maximum absorbance value compensated for background correction as directed for the particular instrument, and draw the best straight line.

Atomize 20 µL of the *Sample Solution* under identical conditions, and measure its corrected maximum absorbance. From the *Standard Curve*, determine the concentration C, in µg/mL, of the *Sample Solution*. Calculate the quantity, in mg/kg, of lead in the sample by the formula:

$$10C/W,$$

in which W is the weight, in g, of the sample taken.

Iodine Value, page 505

Change the penultimate sentence of the paragraph entitled *Wijs Solution* to read:

Alternatively, Wijs solution may be prepared by dissolving 16.5 g of iodine monochloride, ICl, in 1000 mL of glacial acetic acid.

Insert the following new test to precede the test entitled *Sulfur Dioxide*, page 546:

Reducing Sugars Assay

Apparatus Mount a ring support on a ringstand 1 to 2 in. above a gas burner, and mount a second ring 6 to 7 in. above the first. Place a 6-in. open-wire gauze on the lower ring to support a 250-mL Erlenmeyer flask, and place a 4-in. watch glass with a center hole on the upper ring to deflect heat. Attach a 25-mL buret to the ringstand so that the tip just passes through the watch glass centered above the flask. Place an indirectly lighted white surface behind the assembly for observing the endpoint.

Standardized Fehling's Solution Measure a quantity of *Fehling's Solution A*, add an equal quantity of *Fehling's Solution B*, and mix (see *Cupric Tartrate TS, Alkaline*, page 560). Immediately before use, standardize as follows: transfer 3.000 g of primary standard dextrose (NIST Standard Reference Material or equivalent), previously dried in vacuum at 100° for 2 h, into a 500-mL volumetric flask, dissolve in and dilute to volume with water, and mix. Pipet 25 mL of the mixed Fehling's solution into a 200-mL Erlenmeyer flask containing a few glass beads, and titrate with the standard dextrose solution as directed under *Procedure*. Adjust the concentration of *Fehling's Solution A* by dilution or the addition of copper sulfate, so that the titration requires 20.0 mL of the standard dextrose solution.

Procedure Transfer about 3 g of the sample, accurately weighed, into a 500-mL volumetric flask, dissolve in and dilute to volume with water, and mix. Pipet 25.0 mL of *Standardized Fehling's Solution* into a 200-mL Erlenmeyer flask containing a few glass beads, and add the sample solution from a buret to within 0.5 mL of the anticipated endpoint (determined by preliminary titration). Immediately place the flask on the wire gauze of the *Apparatus*, and adjust the burner so that the boiling point will be reached in about 2 min. Bring to a boil, and boil gently for 2 min. As boiling continues, add 2 drops of a 1% aqueous solution of methylene blue, and complete the titration within 1 min by adding the sample solution dropwise or in small increments until the blue color disappears. Record the volume of sample solution required as V, in mL. Calculate the percentage of reducing sugars, as D-glucose on the dried basis, by the formula:

$$(500 \times 0.12 \times 100)/(V \times W),$$

in which W is the weight of the sample, in g, of dry substance.

Replace the *Test* entitled *Sulfur Dioxide*, page 546, with the following:

Sulfur Dioxide Determination
(Based on AOAC Method 962.16)

Reagents

3% Hydrogen Peroxide Solution Dilute 30% hydrogen peroxide to 3% with water. Just before use, add 3 drops of methyl red TS and titrate to a yellow endpoint using 0.01 N sodium hydroxide. If the endpoint is exceeded, discard the solution and prepare another 3% hydrogen peroxide solution.

Standardized Titrant Prepare a solution of 0.01 N sodium hydroxide.

Nitrogen A source of high-purity nitrogen is required with a flow regulator that will maintain a flow of 200 ± 10 mL/min. To guard against the presence of oxygen in the nitrogen, an oxygen scrubbing apparatus or solution such as an alkaline pyrogallol trap may be used. Prepare the pyrogallol trap as follows: add 4.5 g of pyrogallol to the trap, purge the trap with nitrogen for 2 to 3 min, and add potassium hydroxide solution (65 g of potassium hydroxide added to 85 mL of water) to the trap (*Caution*: exothermic reaction) while maintaining an atmosphere of nitrogen in the trap.

Sample Preparation (*for solids*) Transfer 50 g of the sample, or a quantity of the sample with a known quantity of sulfur dioxide (500 to 1500 µg of SO_2), to a food processor or blender, if necessary. Add 50 mL of 5% ethanol in water and briefly grind the mixture, reserving another 50 mL of 5% ethanol in water to

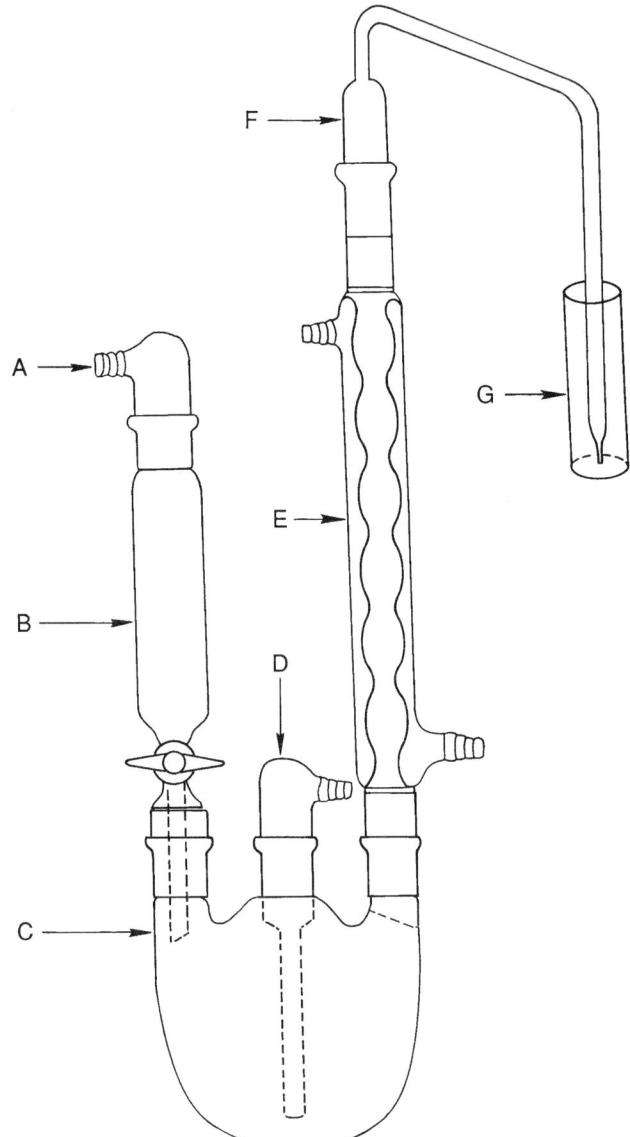

FIGURE 44 The optimized Monier-Williams apparatus. Component identification is given in text (component F is depicted in FIGURE 45).

rinse the blender jar. Grinding or blending should be continued only until the food is chopped into pieces small enough to pass through the 24/40 point of a flask (see Fig. 44).

Sample Preparation (*for liquids*) Mix 50 g of the sample, or a quantity with a known amount of sulfur dioxide (500 to 1500 µg of SO_2), with 100 mL of 5% ethanol in water.

Apparatus The apparatus shown diagrammatically (Fig. 44) is designed to accomplish the selective transfer of sulfur dioxide from the sample in boiling aqueous hydrochloric acid to the *3% Hydrogen Peroxide Solution*. This apparatus is easier to assemble than the official apparatus, and the back pressure inside the apparatus is limited to the unavoidable pressure due to the height of the *3% Hydrogen Peroxide Solution* above the tip of the bubbler, F. Keeping the back pressure as low as possible reduces the likelihood that sulfur dioxide will be lost through leaks. (NOTE: Tygon and silicon tubing should be preboiled before use in this procedure.)

The apparatus should be assembled as shown in Fig. 44, with a thin film of stopcock grease on the sealing surfaces of all the joints except the joint between the separatory funnel and the flask. Each joint should be clamped together to ensure a complete seal throughout the analysis. The separatory funnel, B, should have a capacity of 100 mL or greater. An inlet adapter, A, with a hose connector (Kontes K-183000 or equivalent) is required to provide a means of applying a head of pressure above the solution. (A pressure-equalizing dropping funnel is not recommended because condensate, perhaps with sulfur dioxide, is deposited in the funnel and the side arm.) The round-bottom flask, C, is a 1000-mL flask with three 24/40 tapered joints. The gas inlet tube, D (Kontes K-179000 or equivalent), should be of sufficient length to permit introduction of the nitrogen within 2.5 cm of the bottom of the flask. The Allihn condenser, E (Kontes K-431000-2430 or equivalent), has a jacket length of 300 mm. The bubbler, F, is fabricated from glass according to the dimensions given in Fig. 45, and it has the same dimensions as a 50-mL graduated cylinder (see Fig. 45). The *3% Hydrogen Peroxide Solution* can be contained in a receiving vessel, G, with an inside diameter of about 2.5 cm and a depth of 18 cm.

Buret Use a 10-mL buret with overflow tube and hose connections for an Ascarite tube or equivalent air-scrubbing

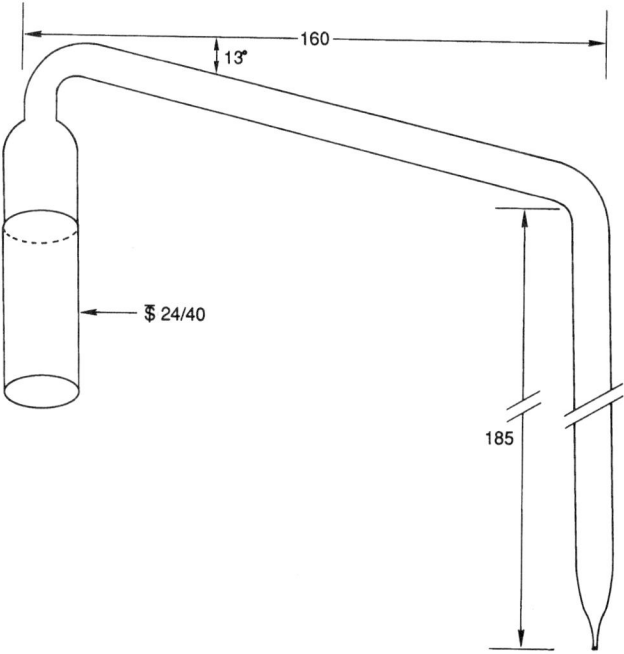

FIGURE 45 Diagram of bubbler (F in FIGURE 44). Lengths are given in mm.

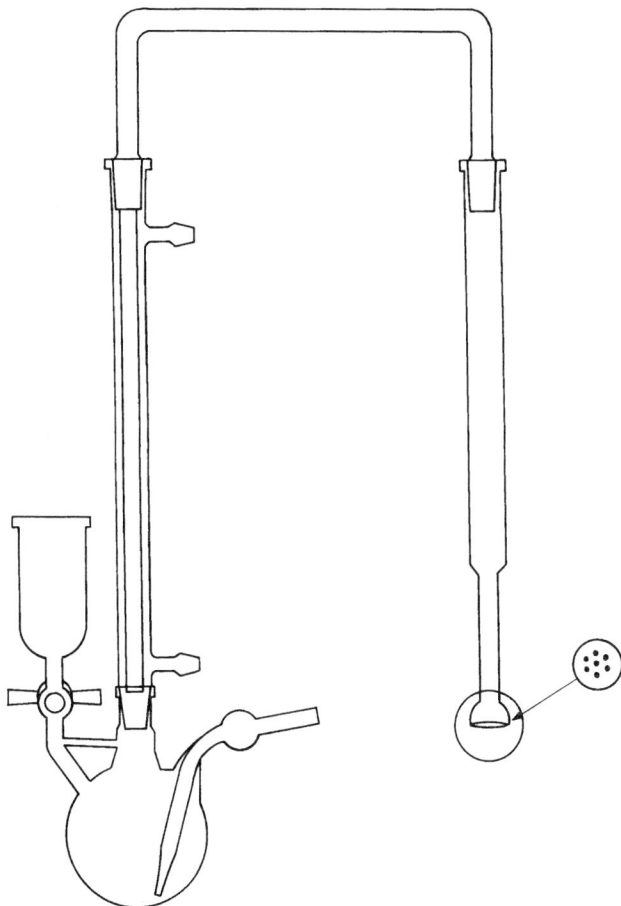

FIGURE 46 Raney nickel reduction apparatus.

apparatus. This will permit the maintenance of a carbon dioxide-free atmosphere over the *Standardized Titrant*.

Chilled Water Circulator The condenser must be chilled with a coolant, such as 20% methanol–water, at a flow rate so that the condenser outlet temperature is maintained at 5°. A circulating pump equivalent to the Neslab Coolflow 33 is suitable.

Determination Assemble the apparatus as shown in Fig. 44. The flask must be positioned in a heating mantle that is controlled by a power-regulating device such as Variac or equivalent. Add 400 mL of distilled water to the flask. Close the stopcock of the separatory funnel, and add 90 mL of 4 N hydrochloric acid to the separatory funnel. Begin the flow of nitrogen at a rate of 200 ± 10 mL/min. The condenser coolant flow must be initiated at this time. Add 30 mL of *3% Hydrogen Peroxide Solution*, which has been titrated to a yellow endpoint with the *Standardized Titrant*, to the receiving vessel, G. After 15 min, the apparatus and the water will be thoroughly deoxygenated, and the apparatus will be ready for sample introduction.

Sample Introduction and Distillation Remove the separatory funnel, and quantitatively transfer the food sample in aqueous ethanol to the flask. Wipe the tapered joint clean with a laboratory tissue, apply stopcock grease to the outer joint of the separatory funnel, and return the separatory funnel to the tapered joint flask. The nitrogen flow through the *3% Hydrogen Peroxide Solution* should resume as soon as the funnel is reinserted into the appropriate joint in the flask. Examine each joint to ensure that it is sealed.

Apply a head pressure above the hydrochloric acid solution in the separatory funnel with a rubber bulb equipped with a valve. Open the stopcock in the separatory funnel, and permit the hydrochloric acid solution to flow into the flask. Continue to maintain sufficient pressure above the acid solution to force the solution into the flask. The stopcock may temporarily be closed, if necessary, to pump up the pressure above the acid. To guard against the escape of sulfur dioxide into the separatory funnel, close the stopcock before the last few mL drain out of the separatory funnel.

Apply the power to the heating mantle. Use a power setting that will cause 80 to 90 drops of condensate to return to the flask from the condenser per min. After 1.75 h of boiling, cool the contents of the 1000-mL flask at the condensation rate stated above, and remove the contents of the receiving vessel, G.

Titration Add 3 drops of *Methyl Red Indicator*, and titrate the above-mentioned contents with the *Standardized Titrant* to a yellow endpoint that persists for at least 20 s. Calculate the sulfur dioxide content, expressed as µg of sulfur dioxide per g of food (µg/g or mg/kg) as follows:

$$\text{mg/kg} = (32.03 \times V_B \times N \times 1000)/Wt,$$

in which 32.03 is the milliequivalent weight of sulfur dioxide, in mg; V_B is the volume, in mL, of the *Standardized Titrant* of normality, N, required to reach the endpoint; the factor 1000 converts mg to µg; and Wt is the weight, in g, of food sample introduced into the 1000-mL flask.

Insert the following table at the end of *Invert Sugar*, Second Supplement, page 84.

International Refractive Index Scale of ICUMSA[1] for Pure Sucrose Solutions at 20°C and 589 nm[2]

Sucrose g/100 g	0.0	0.1	0.2	0.3	0.4	0.5	0.6	0.7	0.8	0.9
56	1.4329	4332	4334	4336	4338	4340	4343	4345	4347	4349
57	1.4352	4354	4356	4358	4360	4363	4365	4367	4369	4372
58	1.4374	4376	4378	4380	4383	4385	4387	4389	4392	4394
59	1.4396	4398	4401	4403	4405	4407	4410	4412	4414	4417
60	1.4419	4421	4423	4426	4428	4430	4432	4435	4437	4439
61	1.4442	4444	4446	4448	4451	4453	4455	4458	4460	4462
62	1.4464	4467	4469	4471	4474	4476	4478	4481	4483	4485
63	1.4488	4490	4492	4495	4497	4499	4502	4504	4506	4509
64	1.4511	4513	4516	4518	4520	4523	4525	4527	4530	4532
65	1.4534	4537	4539	4541	4544	4546	4548	4551	4553	4556
66	1.4558	4560	4563	4565	4567	4570	4572	4575	4577	4579
67	1.4582	4584	4586	4589	4591	4594	4596	4598	4601	4603
68	1.4606	4608	4610	4613	4615	4618	4620	4623	4625	4627
69	1.4630	4632	4635	4637	4639	4642	4644	4647	4649	4652
70	1.4654	4657	4659	4661	4664	4666	4669	4671	4674	4676
71	1.4679	4681	4683	4686	4688	4691	4693	4696	4698	4701
72	1.4703	4706	4708	4711	4713	4716	4718	4721	4723	4726
73	1.4728	4730	4733	4735	4738	4740	4743	4745	4748	4750
74	1.4753	4756	4758	4761	4763	4766	4768	4771	4773	4776
75	1.4778	4781	4783	4786	4788	4791	4793	4796	4798	4801
76	1.4804	4806	4809	4811	4814	4816	4819	4821	4824	4826
77	1.4829	4832	4834	4837	4839	4842	4844	4847	4850	4852
78	1.4855	4857	4860	4862	4865	4868	4870	4873	4875	4878
79	1.4881	4883	4886	4888	4891	4894	4896	4899	4901	4904
80	1.4907	4909	4912	4914	4917	4920	4922	4925	4928	4930
81	1.4933	4935	4938	4941	4943	4946	4949	4951	4954	4957
82	1.4959	4962	4964	4967	4970	4972	4975	4978	4980	4983
83	1.4986	4988	4991	4994	4996	4999	5002	5004	5007	5010
84	1.5012	5015	5018	5020	5023	5026	5029	5031	5034	5037
85	1.5039									

[1] International Commission of Uniform Methods of Sugar Analysis (ICUMSA), 23 Avenue d'Iena, Paris 16 eme, France.
[2] Temperature conditions for refractometric sucrose (dry substance) measurements at 589 nm.

7/ Solutions and Indicators

Replace the *Test Solution* entitled *Cupric Citrate TS, Alkaline*, page 560, with the following:

Cupric Citrate TS, Alkaline (*Benedict's Qualitative Reagent*) With the aid of heat, dissolve 173 g of sodium citrate, $C_6H_5Na_3O_7 \cdot 2H_2O$, and 117 g of sodium carbonate, $Na_2CO_3 \cdot H_2O$, in about 700 mL of water, and filter through paper, if necessary. In a separate container, dissolve 17.3 g of cupric sulfate, $CuSO_4 \cdot 5H_2O$, in about 100 mL of water, and slowly add this solution, with constant stirring, to the first solution. Cool the mixture, dilute to 1000 mL, and mix.

Insert the following *Test Solution* to precede *Hydrochloric Acid*, page 561:

Fuchsin–Sulfurous Acid TS Dissolve 200 mg of basic fuchsin in 120 mL of hot water, and allow the solution to cool. Add a solution of 2 g of anhydrous sodium sulfite in 20 mL of water, and then add 2 mL of hydrochloric acid. Dilute the solution with water to 200 mL, and allow to stand for at least 1 h. Prepare this solution fresh.

Insert the following after *Indicator Papers and Test Papers*, page 570, as a new section entitled *Detector Tubes*:

Detector Tubes

Ammonia Detector Tube A fuse-sealed glass tube (Draeger or equivalent) that is designed to allow gas to be passed through it and that contains suitable absorbing filters and support media for the indicator bromophenol blue. The Draeger Reference Number is CH 20501; the measuring range is 5 to 70 ppm. (NOTE: Suitable detector tubes are available from National Draeger, Inc., P.O. Box 120, Pittsburgh, PA 15205-0120. Tubes other than those specified in the monograph may be used in accordance with the section entitled *Codex Specifications* in the *General Provisions*.)

Carbon Dioxide Detector Tube A fuse-sealed glass tube (Draeger or equivalent) that is designed to allow gas to be passed through it and that contains suitable absorbing filters and support media for the indicators hydrazine and crystal violet. The Draeger Reference Number is CH 30801; the measuring range is 0.01 to 0.30%.

Carbon Monoxide Detector Tube A fuse-sealed glass tube (Draeger or equivalent) that is designed to allow gas to be passed through it and that contains suitable absorbing filters and support media for the indicators iodine pentoxide, selenium dioxide, and fuming sulfuric acid. The Draeger Reference Number is CH 25601; the measuring range is 5 to 150 ppm.

Chlorine Detector Tube A fuse-sealed glass tube (Draeger or equivalent) that is designed to allow gas to be passed through it and that contains suitable absorbing filters and support media for the indicator *o*-toluidine. The Draeger Reference Number is CH 24301; the measuring range is 0.2 to 3 ppm.

Nitric Oxide–Nitrogen Dioxide Detector Tube A fuse-sealed glass tube (Draeger or equivalent) that is designed to allow gas to be passed through it and that contains suitable absorbing filters and support media for an oxidizing layer and the indicator diphenylbenzidine. The Draeger Reference Number is CH 29401; the measuring range is 0.5 to 10 ppm.

Sulfur Dioxide Detector Tube A fuse-sealed glass tube that is designed to allow gas to be passed through it and that contains suitable absorbing filters and support media for an iodine–starch indicator. The Draeger Reference Number is CH 31701; the measuring range is 1 to 25 ppm.

Water Vapor Detector Tube A fuse-sealed glass tube (Draeger or equivalent) that is designed to allow gas to be passed through it and that contains suitable absorbing filters and support media for the indicator, which consists of a selenium sol in suspension in sulfuric acid. The Draeger Reference Number is CH 67 28531; the measuring range is 5 to 200 mg/m^3.

8/ *General Information*

Replace the *Operating Procedures of the* Food Chemicals Codex with the following:

Operating Procedures of the *Food Chemicals Codex*

ORGANIZATION

The *Food Chemicals Codex* project is an activity of the Food and Nutrition Board, under the Institute of Medicine of the National Academy of Sciences. The immediate responsibility for developing the *Food Chemicals Codex* (FCC) lies with the Board's Committee on Food Chemicals Codex. The Committee consists of 12 to 15 members, chosen for their expertise in the various aspects of the Committee's work, who are appointed, upon recommendation of the Food and Nutrition Board and the president of the Institute of Medicine, by the Chairman of the National Research Council. Members are paid no consulting fees or honoraria and are reimbursed only for expenses incurred while attending meetings and other activities of the Committee.

FUNCTIONS OF THE COMMITTEE ON FOOD CHEMICALS CODEX

The principal functions of the Committee on Food Chemicals Codex are as follows:

- To establish the general policies and guidelines by which FCC specifications are prepared.
- To evaluate comments submitted by interested parties on any aspect of the specifications and test procedures.
- To propose means by which the specifications may be kept current in reflecting food-grade quality on the basis of product safety and good manufacturing practice.
- To provide information on issues dealing with specifications for particular substances and analytical test procedures.
- To seek the advice of specialists when additional expert opinion is needed in making decisions regarding the appropriateness of specifications.
- To establish working groups consisting of Committee members and other experts to address specific issues relevant to monograph development and report to the full Committee their findings and recommendations.
- To consider and act upon any other matter concerning the development and publication of specifications and test procedures for food-grade ingredients.
- To approve the final manuscript for review by the National Academy of Sciences before the publication of any edition of the *Food Chemicals Codex* or its supplements.

The business of the Committee is conducted through a central office at the National Academy of Sciences in Washington, D.C. The Academy's staff officer (project director) is responsible for coordinating all of the Committee's activities. The Committee meets in regular session, usually twice a year, to discuss the project's progress, including technical and policy issues relevant to the project. Ad hoc meetings on short-term projects are held as needed and are conducted by one or more members of the Committee and by the project director. Workshops and symposia are organized as appropriate to exchange information with interested parties on key issues, whether of broad or limited scope. (NOTE: Further information concerning the operation of committees of the National Research Council is contained in two pamphlets, single copies of which are available upon request from the Codex office: *Of Questions and Committees: How the National Research Council Does Its Work* and *General Information for Members of Committees of the National Research Council*.)

REQUIREMENTS FOR LISTING SUBSTANCES IN THE *FOOD CHEMICALS CODEX*

The requirements are as follows: (1) the substance is permitted for use in foods or in food processing by the U.S. Food and Drug Administration (or, in certain cases, by other countries in which *Food Chemicals Codex* specifications are recognized), (2) it is commercially available, and (3) suitable specifications and analytical test procedures are available to determine its purity.

CRITERIA FOR *FOOD CHEMICALS CODEX* GRADE

The specifications published in the *Food Chemicals Codex* are based primarily on the criteria of safety and good manufacturing practices (GMP). An FCC-grade substance is one that is prepared under GMP (see page 573) and is of such purity as to ensure that potentially harmful or objectionable contaminants are not present at levels that would represent a hazard to the consumer of the foods in which the substance is intended to be used. Thus, *Food Chemicals Codex* specifications define substances of a level of quality sufficiently high to represent a reasonable certainty of safety when they are used under customary conditions of intentional use in foods or in food processing. The specifications generally represent acceptable levels of quality and purity of food-grade substances available in the United States and in other countries in which *Food Chemicals Codex* specifications are recognized. Because the different types of ingredients used in foods are diverse and complex, few general criteria can be established that will apply to all substances for which *Food Chemicals Codex* specifications are prepared. The Committee recognizes that limits and tests cannot be provided to cover all possible unusual or unexpected impurities, the presence of which would be inconsistent with GMP. This matter is discussed further under *Trace Impurities*, on page 3 of the *General Provisions*, and in *General Good Manufacturing Practice Guidelines for Food Chemicals*, commencing on page 573.

In addition to impurity limits, specifications consist, at the minimum, of the following (where applicable): empirical formula, structural formula, and molecular weight; description of the substance, including physical form, odor, and solubility (see the descriptive terms for solubility on page 4 of the *General Provisions*); identification; assay (or a quantitative test to serve as an assay); as appropriate, such physicochemical characteristics as specific rotation, melting range or solidification point, viscosity, specific gravity, refractive index, pH, etc.; loss on drying or water content; residual solvents; where appropriate, limits for mycotoxins and microbiological contaminants; and limits for by-products and other adventitious constituents usually occurring in, or arising from the manufacture of, the substance. Furthermore, the data provided, taken together, represent a complete compositional analysis of the substance. Also, information is provided on how the substance is to be packaged and stored to maintain its integrity and on its functional use(s) in foods. If the ingredient contains an "added substance," mention is made of this fact to enable the Committee to judge whether the specifications should include it (see *Added Substances* on page 5 of the *General Provisions*).

PROCEDURES FOR SUBMISSION AND DEVELOPMENT OF SPECIFICATIONS

The Committee will consider suggested specifications submitted with supporting data, such as elaborated above, by any interested party, including food ingredient manufacturers and suppliers, food processors, and industry associations. Suggested specifications should be submitted, in duplicate, to Food Chemicals Codex, National Academy of Sciences, 2101 Constitution Avenue, N.W., Washington, DC 20418. Suggested specifications are examined critically by the Committee and/or the project staff and are often expanded to meet the general criteria required by the Committee. Because Committee discussions involving quality characteristics of substances used in food or food processing might result in sharing privileged or proprietary information, such discussions can be requested to be held in closed sessions. The final outcome of such discussions must be openly shared with all manufacturers, users, and parties interested in the substance discussed; therefore, open discussions are the norm, except for unusual circumstances. Where privileged or proprietary information is concerned, the project staff can put such information in a format so that the end results are not associated with particular manufacturers or users. The National Academy of Sciences–Institute of Medicine is not bound by the Freedom of Information Act; thus, information shared with the staff or the Committee need not be made accessible to others. If the submitter wishes, information can be labeled confidential, and the project staff will exercise due discretion. A new monograph drafted by the Committee and/or the staff is then sent to the originator for comment (and to any other manufacturers of that substance that can be identified). After the draft has gone through this process and all necessary revisions have been made, the Committee votes by mail ballot whether to propose these specifications for public comment. If the Committee finds deficiencies, or if any questions are raised, the draft is returned to the originator and other interested parties with the Committee's comments and recommendations for improvement. Once a draft has gained Committee acceptance, availability of the proposed specification for comment is announced in the *Federal Register*. This notification allows the public and other interested parties as well as manufacturers and users that may be inadvertently overlooked to provide their comments to the Committee. Once the public comments are considered and any necessary changes made, the Committee votes to determine whether the monograph is suitable for publication. Monographs are then reviewed by the National Academy of Sciences and, if approved as official monographs, are announced or are published in a supplement or the next edition of the *Food Chemicals Codex*.

PROCEDURE FOR REVISING SPECIFICATIONS

The specifications of the *Food Chemicals Codex* are subject to revision at any time. Suggestions for revision may be initiated by regulatory bodies, manufacturers, suppliers, or users of the ingredients; by the Committee itself; or by any other interested parties. All suggestions for revision must be accompanied by supporting data. In the case of revisions of test procedures and analytical methods, comparative data for both the existing and suggested procedures must be submitted. Where changes in limits or other tolerances are suggested, supporting data should be presented on representative production batches. Suggestions for changing the limits of certain impurities (e.g., arsenic, heavy metals, lead, fluoride, and mercury) may require the submission of safety data and information concerning the daily intake of the substance. All suggestions for revision, together with the supporting data, are reviewed by the Committee on Food Chemicals Codex and/or by the staff. If other manufacturers are involved (and can be identified), they are also asked to comment. If the Committee finds deficiencies, or if any questions arise, the suggested revised specifications are returned to the originator (and other manufacturers, where appropriate) with the Committee's comments or questions. If agreement cannot be reached at this point between the Committee and the originator, or among manufacturers and other interested parties, a special meeting may be held to discuss the matter, or the parties involved may be invited to one of the Committee's regular meetings to examine the question in depth. Approved revisions are published in either the next edition of the *Food Chemicals Codex* or in a supplement by the same procedure described above in *Procedures for Submission and Development of Specifications*.

FURTHER INFORMATION

Users of the *Food Chemicals Codex* should become thoroughly familiar with the *General Provisions* pertaining to this edition (see pages 1–5). Additional information concerning the operation of the project and the revision process will be found in the *Preface to the Third Edition*, beginning on page xvii. Inquiries regarding any aspect of the operation of the *Food Chemicals Codex* project may be directed to Food Chemicals Codex, National Academy of Sciences, 2101 Constitution Avenue, N.W., Washington, DC 20418 (telephone 202-334-2580; facsimile 202-334-2939; electronic bulletin board 202-334-1738).

9/ Infrared Spectra

2-Acetyl Pyrazine

Insert the infrared spectrum of *2-Acetyl Pyrazine* to precede the spectrum for *Almond Oil, Bitter, FFPA*, page 585.

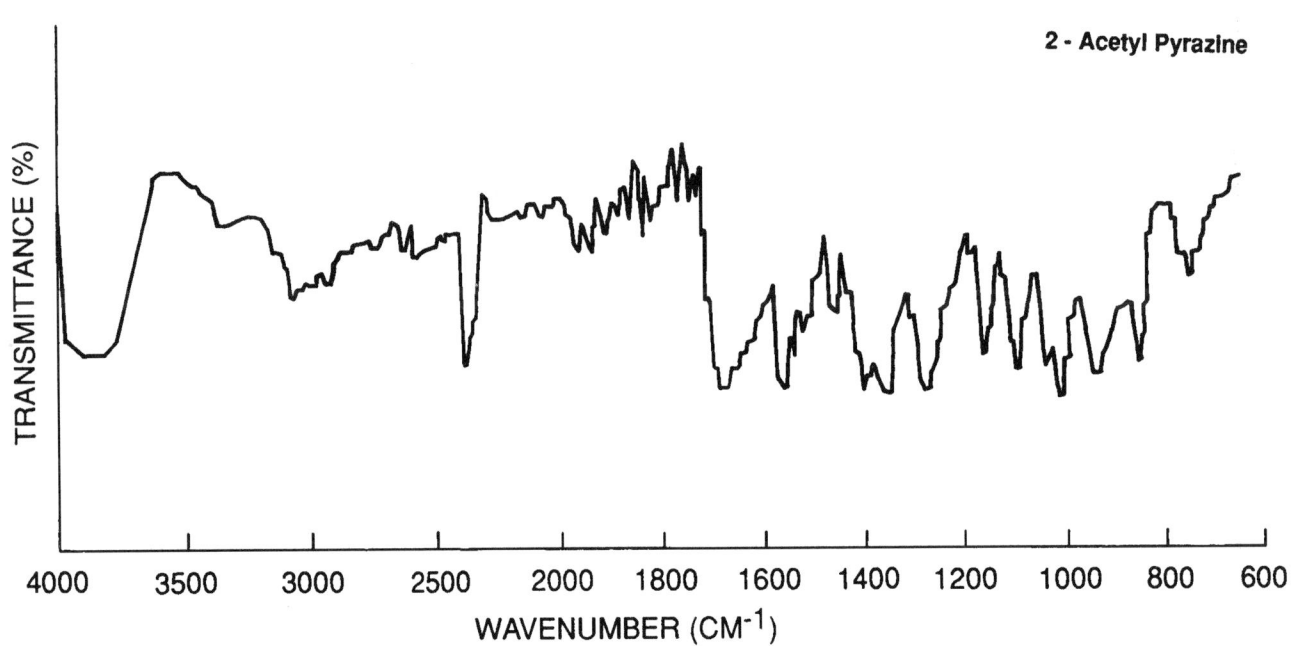

p-Cymene

Insert the infrared spectrum of p-*Cymene* to precede the spectrum for *Diethyl Malonate*, page 682.

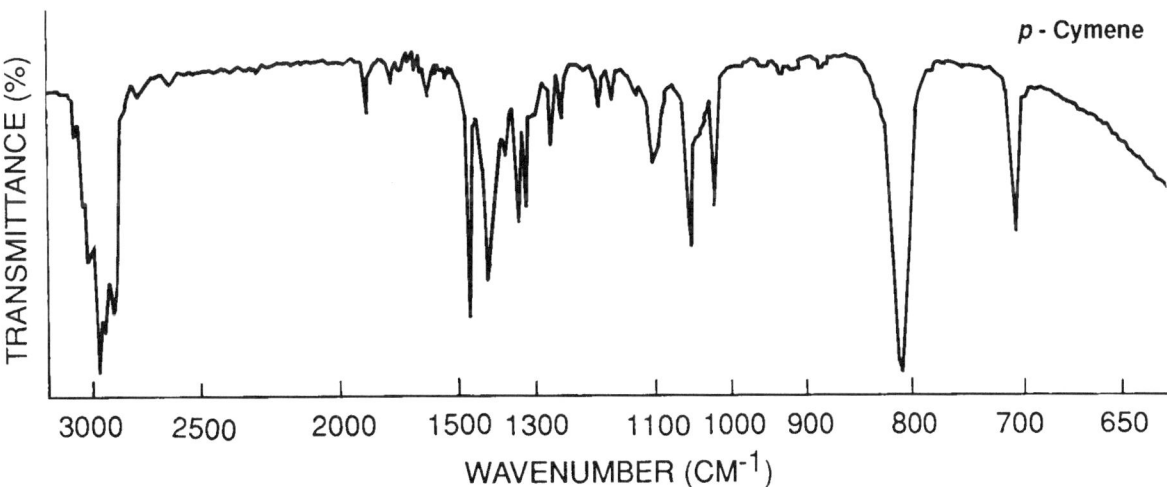

6-Methylcoumarin

Insert the infrared spectrum of *6-Methylcoumarin* to precede the spectrum for *Methyl Eugenol*, page 700.

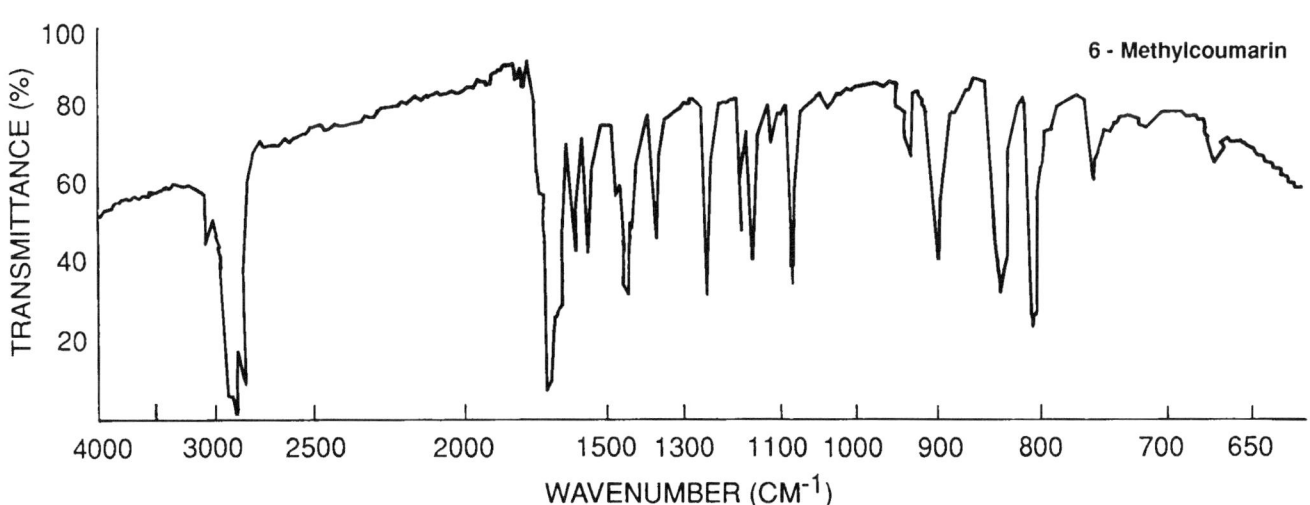

Index

Page citations refer to the First (1–34), Second (35–90), and Third (91–185) Supplements to the Third Edition of the *Food Chemicals Codex*. An asterisk (*) indicates a new listing.

Aceteugenol, 15
3-Acetyl-2,5-dimethyl Furan, 12, 64
Acetyl Eugenol, 15
N-Acetyl-L-Methionine, 97
2-Acetyl Pyrazine, 66, 178
Achilleic Acid, 97
Acid Hydrolyzed Proteins, 3
Aconitic Acid, 97
DL-Alanine, 98
L-Alanine, 99
Alcohol C-6, 17
Alcohol C-8, 19
Alcohol C-10, 14
Alcohol C-12, 17
Aldehyde C-7, 16
Aldehyde C-14, 157
Allura Red AC, 48
p-Allylanisole, 15, 65
Allyl Caproate, 12
4-Allylguaiacol, 15
Allyl Heptanoate, 22
Allyl Heptoate, 22
Allyl Hexanoate, 12
Allyl Ionone, 12
Allyl α-Ionone, 12
4-Allyl-2-methoxyphenol, 15
4-Allyl-2-methoxyphenyl Acetate, 15
Aluminum Ammonium Sulfate, 4
Aluminum Potassium Sulfate, 4
Aluminum Sodium Sulfate, 5, 99

Aluminum Sulfate, 5
Ammonium Bicarbonate, 5
Amyl Butyrate, 17
Amylcinnamaldehyde, 12
α-Amylcinnamaldehyde, 12
Amyl Heptanoate, 66
Amyl Salicylate, 17
*Amyl Vinyl Carbinol, 162
Anethole, 64
Angelica Seed, Oleoresin, 9
Anhydrous Calcium Chloride, 38
p-Anisaldehyde, 18
Anise, Oleoresin, 10
Anisic Aldehyde, 18
Anisyl Acetate, 12
Anisylacetone, 66
*Anisyl Formate, 158
Annatto Extracts, 37, 99
L-Arginine, 99
L-Arginine Monohydrochloride, 99
L-Asparagine, 99
Aspartame, 5, 31
DL-Aspartic Acid, 100
L-Aspartic Acid, 100
Assay by Determination of Aldehydes and Ketones–Hydroxylamine Method, 24
Assay by Determination of Esters, 24
Assay by Determination of Esters, High-Boiling Method, 24

Assay by Gas-Liquid Chromatography, 25
Assay for Determination of Phenols, 25
*Atomic Absorption Spectrophotometry (Graphite Furnace Method), 168
Autolyzed Yeast Extract, 5
BHA, 38
Barium Hydroxide, 88
Basil, Oleoresin, 10
Bay Oil, 6
Beet Sugar, 149
Benzaldehyde, 13
Benzaldehyde Glyceryl Acetal, 66
Benzene (in Paraffinic Hydrocarbon Solvents), 71
Benzodihydropyrone, 22
Benzyl Acetate, 13
Benzyl Butyrate, 26
Benzyl Cinnamate, 13, 156
*Benzyl Formate, 158
Benzyl Isobutyrate, 156
Benzyl Isovalerate, 156
Benzyl Propionate, 26
Benzyl Salicylate, 13
Black Pepper Oil, 100
Brilliant Blue FCF, 43
Brominated Vegetable Oil, 38
*Butane, 101
*n-Butane, 101

1-Butanol, 64
Butter Starter Distillate, 148
Butyl Acetate, 13
n-Butyl Acetate, 13
Butyl Alcohol, 64
Butyl Aldehyde, 13
Butylated Hydroxyanisole, 38
Butyl Isovalerate, 22
*Butyl Octadecanoate, 158
Butyl Phenylacetate, 66
*Butyl Stearate, 158
Butyraldehyde, 13
Calcium Carbonate, 38
Calcium Chloride, 102
Calcium Chloride, Anhydrous, 38
    Acid-Insoluble Matter in, 70
Calcium Gluconate, 102
Calcium Oxide, 6
Calcium Pantothenate, 103
Calcium Pantothenate, Calcium
    Chloride Double Salt, 103
Calcium Pantothenate, Racemic, 103
*Calcium Sorbate, 104
Calcium Sulfate, 39
Camphene, 13
*D-Camphor, 158
Cane Sugar, 149
Canola Oil, 104
Caproic Aldehyde, 16
Capryl Alcohol, 19
Caraway, Oleoresin, 10
Carbamide, 152
Carbon, Activated, 6, 106
Cardamom, Oleoresin, 10
Carmine, 6
Carnauba Wax, 107
Carrageenan, 39
Carrot Seed Oil, 107
Carvacrol, 13
*L-Carveol, 158
*L-Carvyl Acetate, 158
β-Caryophyllene, 13
Casein and Caseinate Salts, 39
Celery Seed Oil, 6
Chromatography, 27, 73
Cinnamal, 13
Cinnamaldehyde, 13
Cinnamic Aldehyde, 13
Cinnamyl Acetate, 14
*Cinnamyl Butyrate, 158
*Cinnamyl Cinnamate, 158
Cinnamyl Formate, 14, 64
*Cinnamyl Isobutyrate, 158
Cinnamyl Propionate, 64
Citric Acid, 107

Citridic Acid, 97
Citronellal, 64
Citronellol, 14
Citronellyl Acetate, 14
Citronellyl Formate, 14
Coconut Oil, 40
Colors, FD&C (also see specific color
    under FD&C), 73
  Chromium Test, 73
  Ether Extracts Test, 74
  Leuco Base Test, 74
  Mercury Test, 75
  Side Reaction Products Test, 77
  Sodium Chloride Test, 76
  Sodium Sulfate Test, 76
  Total Color Test, 76
  Uncombined Intermediates Test, 77
  Volatile Matter Test, 80
  Water-Insoluble Matter Test, 80
*Copper Sulfate, 108
Coriander, Oleoresin, 10
Corn Oil, 41
*Corn Syrup, 112
Corn Syrup, High-Fructose, 51
Cottonseed Oil, 42
Cubeb, Oleoresin, 10
Cumaldehyde, 14, 64
Cumin, Oleoresin, 10
Cuminal, 14, 64
Cuminic Aldehyde, 14, 64
p-Cuminic Aldehyde, 14, 64
Cupric Citrate TS, Alkaline, 173
*Cupric Sulfate, 108
Cyclamen Aldehyde, 14, 65
p-Cymene, 66, 179
L-Cysteine Monohydrochloride, 109
L-Cystine, 109
Δ-Decalactone, 14, 65
*γ-Decalactone, 158
1-Decanol, Natural, 14
cis-4-Decen-1-al, 65
trans-2-Decen-1-al, 65
Decyl Alcohol, 14
Detector Tubes, 173
Dextrose, 109
Diatomaceous Earth, 110
Dibenzyl Ether, 22
Diethyl Succinate, 15
Dillseed, Oleoresin, 10
1,2-Dimethoxy-4-allylbenzene, 19
2,5-Dimethyl-3-acetylfuran, 12, 64
Dimethyl Benzyl Carbinyl Butyrate, 15
2,6-Dimethyl-5-heptenal, 66, 156
3,7-Dimethyl-2,6-octadien-1-yl Ac-
    etate, 16

*cis-3,7-Dimethyl-2,6-octadien-1-yl
    Acetate, 162
3,7-Dimethyl-2,6-octadien-1-yl Ben-
    zoate, 16
3,7-Dimethyl-2,6-octadien-1-yl Bu-
    tyrate, 16
3,7-Dimethyl-2,6-octadien-1-yl For-
    mate, 16
3,7-Dimethyl-2,6-octadien-1-yl
    Phenylacetate, 16
3,7-Dimethyl-2,6-octadien-3-yl Propi-
    onate, 18
3,7-Dimethyl-3-octanol, 20
3,7-Dimethyl-6-octen-1-al, 14
3,7-Dimethyl-6-octen-1-ol, 14
3,7-Dimethyl-6-octen-1-yl Acetate, 14
3,7-Dimethyl-6-octen-1-yl Formate, 14
α,α-Dimethylphenethyl Butyrate, 15
Dimethylpolysiloxane, 43
*Dimethyl Sulfide, 158
Disodium Guanylate, 6
Δ-Dodecalactone, 15, 65
*γ-Dodecalactone, 158
1-Dodecanol, 17
*Dried Glucose Syrup, 112
Enocianina, 7
Enzyme Preparations, 6
Epsom Salt, 8
Equisetic Acid, 97
Erythrosine, 46
Estragole, 15, 65, 156
Ethoxyquin, 110
Ethyl Butyl Ketone, 16
Ethyl Cinnamate, 15
*Ethylene Brassylate, 160
2-Ethyl Fenchol, 65
Ethyl Formate, 156
*4-Ethylguaiacol, 160
Ethyl Isobutyrate, 22
Ethyl 2-Methylbutyrate, 15
*Ethyl-3-Methylthiopropionate, 160
Ethyl Myristate, 22
*Ethyl 9-Octadecenoate, 160
*Ethyl Oleate, 160
Ethyl 3-Phenylpropenoate, 15
Eugenic Acid, 15
Eugenol, 15
Eugenol Acetate, 15
Eugenyl Acetate, 15
Eugenyl Methyl Ether, 19
Extractable Organic Compounds (in
    Hydrochloric Acid), 80
FCC Operating Procedures, 175
  Criteria for *Food Chemicals Codex*
    Grade, 176

Functions of the Committee on Food Chemicals Codex, 175
    Further Information, 177
    Organization, 175
    Procedure for Revising Specifications, 177
    Procedures for Submission and Development of Specifications, 176
    Requirements for Listing Substances in the *Food Chemicals Codex*, 176
FD&C Blue No. 1, 43
FD&C Blue No. 2, 44
FD&C Green No. 3, 45
FD&C Red No. 3, 46
FD&C Red No. 40, 48
FD&C Yellow No. 5, 49
FD&C Yellow No. 6, 50
Farnesol, 15
Fast Green FCF, 45
Fats and Related Substances, 82
    Cold Test, 82
    Fatty Acid Composition, 82
    Melting Range, 82
    Stability Test, 83
Fennel Oil, 110
Fennel, Oleoresin, 10
Ferrous Fumarate, 110
Fructose, 7, 110
Fructose Corn Syrup, 51
Fuchsin–Sulfurous Acid TS, 173
Fully Hydrogenated Rapeseed Oil, 140
Fusel Oil Refined, 66
Garlic Oil, 111
Gas Chromatographic Analysis of Butyl and Isobutyl Alcohols, 68
Gas Chromatography, 27
Gas Chromatography, Calculation of Response Factors, 73
Gellan Gum, 111
Geranyl Acetate, 16
Geranyl Benzoate, 16
Geranyl Butyrate, 16
Geranyl Formate, 16
Geranyl Phenylacetate, 16
*Glucose Syrup, 112
*Glucose Syrup, Dried, 112
*Glucose Syrup Solids, 112
L-Glutamic Acid, 113
L-Glutamic Acid Hydrochloride, 113
L-Glutamine, 114
*Glutaral, 114
*Glutaraldehyde, 114
Glyceryl Behenate, 115
Glyceryl Monostearate, 116

Glyceryl Tribehenate, 115
Glyceryl Tridocosanoate, 115
*Glyceryl Tripropanoate, 160
Glyceryl-Lacto Esters of Fatty Acids, 122
Glycine, 118
Granulated Sugar, 149
Grape Skin Extract, 7
Gum Ghatti, 118
*Helium, 119
Heptaldehyde, 16
Heptanal, 16
2-Heptanone, 16
3-Heptanone, 16
*2,4-Hexadienoic Acid, Calcium Salt, 104
*Hexahydropyridine, 162
*γ-Hexalactone, 160
Hexaldehyde, 16
Hexanal, 16
Hexanes, 50
1-Hexanol, 17
cis-3-Hexen-1-ol, 16
cis-3-Hexenyl Acetate, 160
**trans*-2-Hexenyl Acetate, 160
n-Hexyl Acetate, 22
Hexyl Alcohol, Natural, 17
Hexyl Isovalerate, 17
High-Fructose Corn Syrup, 51, 119
    Solids in, 84
L-Histidine, 120
L-Histidine Monohydrochloride, 120
Hydrocarbons, Mixed Paraffinic, 50
Hydrochloric Acid, 52
    Extractable Organic Compounds in, 80
Hydrolyzed Milk Protein, 3
Hydrolyzed Plant Protein (HPP), 3
Hydrolyzed Vegetable Protein (HVP), 3
2-Hydroxy-Propanoic Acid Monopotassium Salt, 144
2-Hydroxy-Propanoic Acid Monosodium Salt, 143
*4-Hydroxydecanoic Acid Lactone, 158
*4-Hydroxy-2,5-dimethyl-3(2H)furanone, 160
*4-Hydroxydodecanoic Acid Lactone, 158
*4-Hydroxyhexanoic Acid Lactone, 160
*4-Hydroxy-3-methoxy-ethylbenzene, 160
5-Hydroxynonanoic Acid, Lactone, 66
5-Hydroxyoctanoic Acid, Lactone, 66
4-(*p*-Hydroxyphenyl)-2-butanone, 22

Indian Gum, 118
Indigo Carmine, 44
Indigotine, 44
Indigotine Disulfonate, 44
Invert Sugar, 53, 84, 120, 172
Invert Sugar Syrup, 53
Iodine Value, 169
α-Ionone, 17
β-Ionone, 17
Isoamyl Benzoate, 17, 66
Isoamyl Butyrate, 17, 156
Isoamyl Formate, 26, 156
Isoamyl Salicylate, 17
*Isoborneol, 160
Isobornyl Acetate, 65
*Isobutane, 121
Isobutyl Acetate, 17
Isobutyl Alcohol, 17, 65
Isobutyric Acid, 17
DL-Isoleucine, 121
L-Isoleucine, 7, 121
Isophenylformic Acid, 17
p-Isopropylbenzaldehyde, 14, 64
Isovaleric Acid, 156
Konjac, 122
Konjac Flour, 122
Konjac Gum, 122
Konnyaku, 122
Lactated Mono-Diglycerides, 122
Lactic Acid, 123
Lactose, 85
Lard, 54
Laurel Leaf, Oleoresin, 10
Lauryl Alcohol, Natural, 17
LEAR, 104
DL-Leucine, 123
L-Leucine, 7, 124
Linalyl Propionate, 18
Low Erucic Acid Rapeseed Oil, 104
L-Lysine Monohydrochloride, 124
Magnesium Chloride, 124
Magnesium Oxide, 7
Magnesium Sulfate, 8
*Maltodextrin, 125
Mandarin Oil, Coldpressed, 8
Marjoram, Oleoresin, 10
p-Mentha-6,8-dien-2-ol, 158
p-Mentha-6,8-dien-2-yl Acetate, 158
3-*p*-Menthanol, 18
l-*p*-Menthan-3-one, 18
dl-*p*-Menthan-3-yl Acetate, 18
l-*p*-Menthan-3-yl Acetate, 18
Menth-1-en-8-ol, 20, 65
p-Menth-1-en-8-ol, 65
Menthol, 18

l-Menthone, 18
dl-Menthyl Acetate, 18
l-Menthyl Acetate, 18
*Methional, 162
DL-Methionine, 125
L-Methionine, 125
p-Methoxybenzaldehyde, 18
p-Methoxybenzyl Acetate, 12
*p-Methoxybenzyl Formate, 158
4-p-Methoxyphenyl-2-butanone, 66
2-Methoxypyrazine, 18
4'-Methyl Acetophenone, 18
Methyl Amyl Ketone, 16
*Methyl Benzaldehyde, 162
*p-Methylbenzaldehyde, 164
Methyl Benzoate, 90
2-Methylbutyl Isovalerate, 19, 65
2-Methylbutyl-3-methylbutanoate, 19, 65
α-Methylcinnamaldehyde, 19
6-Methylcoumarin, 66, 179
Methyl Eugenol, 19
Methyl Formate, 55
*Methyl Heptyl Ketone, 162
Methyl Hexyl Ketone, 66
2-Methyl-3-(p-isopropylphenyl)-propionaldehyde, 14, 65
Methyl 2-Methylbutanoate, 19
Methyl 2-Methylbutyrate, 19
*Methyl Nonyl Ketone, 164
*2-Methylpentanoic Acid, 160
*4-Methylpentanoic Acid, 160
*2-Methyl-2-pentenoic Acid, 160
2-Methyl Propanoic Acid, 17
Methyl Salicylate, 19
*Methyl Sulfide Thiobismethane, 158
*3-Methylthiopropionaldehyde, 162
Methyl p-Tolyl Ketone, 18
Mixed Paraffinic Hydrocarbons, 50
Monoammonium L-Glutamate, 126
Monopotassium L-Glutamate, 126
Monosodium L-Glutamate, 126
Monostearin, 116
*Myristaldehyde, 162
Natamycin, 126
*Neryl Acetate, 162
*Nitrogen, 128
*Nitrogen Enriched Air, 128
*Nitrogen Oxide, 129
*Nitrous Oxide, 129
Δ-Nonalactone, 66
*2-Nonanone, 162
Δ-Octalactone, 66
1-Octanol, Natural, 19
2-Octanone, 66

*1-Octene-3-ol, 162
1-Octen-3-yl Acetate, 19
1-Octen-3-yl Butyrate, 19
3-Octyl Acetate, 19
Octyl Alcohol, 19
Octyl Formate, 19
*Octyl Isobutyrate, 162
*Octyl 2-Methylpropanoate, 162
Oleoresins, Spice, 9, 10
Operating Procedures, FCC, 175
    Criteria for *Food Chemicals Codex* Grade, 176
    Functions of the Committee on Food Chemicals Codex, 175
    Further Information, 177
    Organization, 175
    Procedure for Revising Specifications, 177
    Procedures for Submission and Development of Specifications, 176
    Requirements for Listing Substances in the *Food Chemicals Codex*, 176
Origanum, Oleoresin, 10
Ox Bile Extract, 130
Ozone, 131
Palmarosa Oil, 132
Palm Kernel Oil, 55
Palm Oil, 56
Paraffinic Hydrocarbons, Mixed, 50
Parsley Leaf, Oleoresin, 10
Parsley Seed, Oleoresin, 10
Peanut Oil, 57
Pectins, 132
*1,5-Pentanedial, 114
Peppermint Oil, 8
Perlite, 135
Petroleum Wax, Synthetic, 8
Phenethyl Isovalerate, 19
2-Phenethyl 2-Methylbutyrate, 20
Phenoxyethyl Isobutyrate, 20
Phenylacetaldehyde, 20
DL-Phenylalanine, 135
L-Phenylalanine, 135
*Phenylethyl Anthranilate, 162
*Phenylethyl Butyrate, 162
Pimaricin, 126
Pimenta Berries, Oleoresin, 10
α-Pinene, 157
β-Pinene, 157
*Piperidine, 162
Poloxamer 331, 57
Poloxamer 407, 57
Polydextrose, 57, 136
Polydextrose Solution, 59
Potassium Alginate, 8

*Potassium Benzoate, 136
Potassium Bicarbonate, 8, 137
Potassium Carbonate, 8
Potassium Chloride, 137
Potassium Lactate Solution, 137
Potassium Nitrate, 9
Potassium Sorbate, 9, 139
L-Proline, 139
*Propane, 140
1,2,3-Propanetriol Octadecanoate, 116
p-Propenylanisole, 64
Purified Oxgall, 130
Rapeseed Oil, Fully Hydrogenated, 140
Rapeseed Oil, Superglycerinated, 141
*Reducing Sugars Assay, 169
Response Factors in Gas Chromatography, 73
Rhodinyl Acetate, 20
Safflower Oil, 60
DL-Serine, 142
L-Serine, 9, 142
Silicon Dioxide, 9
Sodium Alginate, 9
Sodium Aluminosilicate, 143
Sodium Bicarbonate, 9
Sodium Carbonate, 9
Sodium Choleate, 130
Sodium Lactate Solution, 144
*Sodium Magnesium Aluminosilicate, 145
Sodium Saccharin, 9
Sodium Stearyl Fumarate, 147
Sodium Thiosulfate, 88
Solutions and Indicators, 29
Sorbitol, 147
Sorbitol Solution, 148
Soybean Oil, 61
Spice Oleoresins, 9, 148
Spike Lavender Oil, 10
Starter Distillate, 149
Sucrose, 149
Sugar, 149
Sulfur (by Oxidative Microcoulometry), 85
Sulfuric Acid, 151
Sulfur Dioxide Determination, 170
Sunflower Oil, 62
Sunset Yellow FCF, 50
Superglycerinated Fully Hydrogenated Rapeseed Oil, 141
System Suitability Tests, GLC, 27
TBHQ, 63

Tallow, 63
Tartrazine, 49
Terpineol, 20, 65
Test for Free Phenols, 25
Test for Phenols Using Cassia Flask Method, 25
Test Methods for Flavor Aromatic Chemicals and Isolates, 24
Test Solutions (TS) and Other Reagents, 29
*Tetradecanal, 162
*Tetrahydrofurfuryl Alcohol, 162
Tetrahydrolinalool, 20
Thiamine Mononitrate, 32
L-Threonine, 151
Thyme, Oleoresin, 10
*Thymol, 162

d-α-Tocopheryl Acetate Concentrate, 10
*Tolualdehyde (mixed isomers), 162
*$para$-Tolualdehyde, 164
α-Toluic Aldehyde, 20
Tolyl Acetate, 18
*p-Tolyl Aldehyde, 164
*Tolyl Aldehyde (mixed isomers), 162
Triacetin, 11
Triatomic Oxygen, 131
Triethyl Citrate, 11, 63
*Trimethylamine, 164
4(2,6,6-Trimethyl-1-cyclohexenyl)-3-butene-2-one, 17
4(2,6,6-Trimethyl-2-cyclohexenyl)-3-butene-2-one, 17
3,7,11-Trimethyl-2,6,10-dodecatrien-1-ol, 15

*Tripropionin, 160
DL-Tryptophan, 151
L-Tryptophan, 152
L-Tyrosine, 152
γ-Undecalactone, 157
*2-Undecanone, 164
*Urea, 152
*Valeraldehyde, 164
γ-Valerolactone, 20
L-Valine, 153
Xanthan Gum, 153
Xylitol, 11, 153
*Zein, 154
Zinc Gluconate, 155
Zinc Sulfate, 11